...e
... ... truly

... ...son, Roy J. Lincourt Jr. and h.. ... ...Patricia of Wanakena; a brother, Harold Lincourt of Tucson, AZ; two sisters, Eva Oakes of Richfield Springs, Lena Martin of CT; a grandson, John Lincourt of Melbourne Beach, FL, several nieces, nephews and his companion Marian Rich of Cherry Valley. Besides his parents and wife Angela he was preceded in death by two brothers, Michael and Francis Lincourt and a sister; Charlotte 'lement.

Funeral from *J. Seaton McGrath 'eral Home*, 40 West James Street, ...field Springs on Wednesday at 10:30 u. A Mass of Christian Burial will be ...lebrated in St. Joseph's Church Richfield ...prings at 11:00 AM. The Reverend 'erence Healy will officiate. Interment ...ill follow in St. Joseph's Cemetery in ...uburn.

The family will be present in the funeral ...ome to receive relatives and friends on Tuesday from 2:00 to 4:00 and 6:00 to 8:00 ...m.

SURVEY DRAFTING

*DRAFTING PRACTICES IN
SURVEYING & ENGINEERING
OFFICES*

by

Gurdon H. Wattles

Second Edition

©Copyright 1977 & 1981
by Gurdon H. Wattles

Library of Congress Card
Nos. 77-83530 & 81-52885

ISBN 0-9606962-0-2

Published by:
Wattles Publications
P.O. Box 1706
Tustin, California 92681-1706

9876543

II

Dedication

The lessons in this book are dedicated to the many individuals who will be an integral part of the total effort that goes into the development of land projects — whether it be a small subdivision for homes, an industrial complex, a shopping center or a transcontinental right-of-way.

It is the author's desire that through this instruction a broader knowledge and understanding may be acquired by those who try hard and want so much to learn the techniques of this special drafting for survey work and the appurtenant operations that go with them including plans and profiles with their necessary cross-section and earthwork calculations.

Introduction

Colleges, trade schools and high schools across this land provide numerous courses in the fields of Architectural and Mechanical drafting. Few provide courses on training to those whose interests lie in the fields of Civil or Surveying drafting.

SURVEY DRAFTING by Gurdon H. Wattles, is a thorough, comprehensive textbook on the elements which comprise the complex subject of survey drafting which should, so to speak, fill the gap.

The book begins with a chapter reviewing some examples of field notes and their interpretations which of course is basic to the subject.

Mr. Wattles then leads us through chapters discussing basic plotting of topography and boundaries, relating maps to descriptions, lettering and line work and drafting tools. All of this is laying the ground work for the succeeding chapters wherein he treats with the more complex problem of preparing the actual "working drawings," the topographic map and its related grading plans, cross sections, quantities and earthwork calculations.

The twelfth chapter covers the subject of Special Maps, the modern subdivision map and plans, the map required by the American Land Title Association, court case exhibits and right of way maps. The final chapters deal with the subject of Geographic Control, as it relates to mapping, and with the subject of computerized mapping as it exists today.

All chapters include fine examples of the subject under discussion with typical problems and their solutions. I believe this book provides a needed textbook for a college level course in the specialized area of survey drafting and mapping and at the same time provides a technical tool for use by the draftsman in the work-a-day world who wants to improve the product he produces.

> George D. Shambeck
> Licensed Land Surveyor No. 3419,
>   State of California
> Vice President, Hall & Foreman, Inc.
> Secretary-Treasurer, Orange County Chapter
>   California Council Civil Engineers
>   and Land Surveyors
> Governor, Board of Governors, Southern
>   California Association of Civil
>   Engineers and Land Surveyors
> Past-Chairman and Member, Orange County Area
>   Surveyors' Joint Apprenticeship Committee

Preface

The spectrum of drawings used in surveying by surveyors and engineers for the study and development of physical projects includes land survey maps, topographic maps, orthophoto maps, hydrographic maps, cadastral maps, cross section diagrams, plan-and-profile drawings and, in some cases, graphical interpolations or adaptations.

More specifically, land survey maps run the gamut from a single line for a right-of-way to a closed boundary of one parcel of land to a layout of many parcels or lots designed into groups or blocks with serving streets, alleys and walks.

Topographic maps are made to show vertical as well as horizontal conditions (known as relief), man-made physical improvements (known as culture), water conditions (known as hydrographic) and sometimes, but not always, trees, meadows and agriculture (known as vegetation); this type of map may or may not include surveys of ownership lines.

Orthophoto maps are the result of superimposing, by drawing, property lines, r/w lines etc onto air photos. This combines relief, culture and vegetation with ownership areas but usually without contours.

Hydrographic maps show principally vertical conditions concerned with bodies of water such as ocean beds, lakes, streams, etc., plus horizontal relationships to shorelines, piers, dams, etc.

Cadastral maps cover city or township, county, parish or borough areas and are compiled from public records such as maps and deeds, as well as from field survey information.

Plan-and-profile drawings are developed with the help of cross-section diagrams to guide and control the physical construction of roads, storm drains and flood control channels, sewers, limited access highways, railroads, canals, etc.

Graphical analysis is applied in those areas where it expedites particular demands of the work.

"Cartography" is considered to be the translation of ideas and information relative to maps and charts into diagrammatic form on paper, but that word is not generally implied to represent *all* of the phases of drawing hereinbefore mentioned. According to Webster, "cartography" is "the art or business of drawing or making charts or maps." However, in the field of surveying and engineering, there are several operations other than maps per se which have an association with and are necessary adjuncts to this phase of the profession. These include surveys of property boundaries and right-of-way alignments which in turn require a knowledge of legal descriptions and the production of field notes. Manifestly the transition from conditions in their natural form to engineered design is made possible by the draftsman through the media of maps, plan-and-profile drawings based upon survey field notes, earth work diagramming, calculations and analysis.

Preliminary to the preparation of these drawings, the ideas are often roughly sketched on scratch paper first with side notes for details. Sometimes the field notes of the surveyor are the directives of both the boundary lines and interior information for the drawing, but in many cases, the legal description on the deed dictates the boundary lines of the drawing within which the development takes place.

In the case of a new subdivision, the map reflects the design of the streets and lots as laid out by the engineer but prior to any such design, the topography must be drawn on paper, the cut and fill of dirt must be analyzed, calculated and then translated into plan-and-profile drawings in order to harmonize the streets and lots into the contour of the land.

Conjunctive with all of this drawing, two other accomplishments are desirable: an acquaintance with a certain amount of mathematics and the ability to present the facts, the picture if you will, in understandable language which means not only the accepted forms of delineation in drafting but also good, legible lettering.

Although Webster states, a "cartographer" is "one who makes charts or maps," the present day application requires that such a one must do more. He must gather all the information available — survey notes, previous maps, descriptions, etc. — and put all of it into an understandable graphical display that relates those true existing conditions. He must use ingenuity to bring forth the best readable product by combining the information obtained from the many sources with his experience and knowledge of the standards and techniques of drawing that are applicable to the requirements of the project at hand.

Thus we see the subject here is very broad in scope and to become proficient in producing this specialized kind of drafting requires knowledge of its many facets and even more, much practice in the art.

Hours are devoted to training people in the specialties of surveying and engineering drafting. Many office managers have expressed the wish that there was some source of material and instruction under one cover which would disseminate this special knowledge and save them the necessity of spending much of their time, which represents money, for such training.

Realizing this predicament and responding to the call, the author has endeavored to include those background and operational fundamentals which are necessary to such a preparation along with a cross-section of the various types of drawings. The assumption is made that the user of this book has had at least primary mechanical drawing and mathematics through trigonometry.

The Author

With a background of 45 years' experience in land surveying, civil engineering, architecture and construction work, along with teaching in these areas, Gurdon Wattles is well qualified to discourse on these subjects. He has seen the need for a reference book as well as a text which would be of great value to all who are similarly concerned in understanding the fundamentals of knowledge in this field.

The author's requested participation in the Technical Curricular Programming of several colleges speaks well for his comprehension of teaching application and their appreciation of his technical experience.

The success of Mr. Wattles' fast-selling book, Writing Legal Descriptions, is further criterion and recognition of his vast knowledge in the area of land development.

Besides teaching in private and collegiate classes, the author has conducted many seminars and workshops at the request of professional, governmental and private industry groups.

He has numerous articles published in national magazines and other media, besides which he holds copyrights on maps, books and brochures.

Mr. Wattles, a Fellow in ASCE, has been a member of the Committee on Land Surveying in the Division of Surveying and Mapping of the American Society of Civil Engineers for seven years and was its National Chairman, 1969-1970; he was also National Chairman of the Committee for Improvement of Land Title Records of the American Congress on Surveying and Mapping, 1970-1972, and a member of that Committee seven years; he has been a member of the American Right of Way Association twenty years; as a member of the Planning Committee of the American Bar Foundation's preparation for a conference on Compatible Land Identifiers, he subsequently participated in its 1972 Conference. Mr. Wattles is also Past President of the Orange County Engineering Council and First Past Chairman of the Surveying and Mapping Technical Group of the Los Angeles Section, American Society of Civil Engineers.

The author recently retired from his position as Assistant Vice President and Consulting Title Engineer of Title Insurance and Trust Company and Pioneer National Title Insurance Company. He is a Registered Land Surveyor in California and Nevada. He is listed in "Who's Who" and reported in "Men of Achievement" International and "Notable Americans Directory."

Acknowledgements

The encouragement of others involved in this area of practice has been the motivation in producing this text. Special acknowledgement goes to Professor Richard E. Hauck, Licensed Land Surveyor and Instructor of Surveying at Pasadena City College, and Mr. James W. Robinson, Licensed Land Surveyor, Assistant Vice President, Title Insurance and Trust Company, for their contributions of materials and suggestions. To Mr. Blaise J. Subbiondo, Registered Civil Engineer, Licensed Building Contractor, Educator, for not only historical and current educational support, but also for his inspiration. To Mr. Porter W. McDonnell, Jr., Professional Engineer and Registered Surveyor and Associate Professor of Engineering, The Pennsylvania State University, for his encouragement. To Mr. Jacob F. Rems for his excellent rendering of the illustrations. To Mr. Hugh Halderman, Civil Engineer, for his review and suggestions. To Mr. Ed Fisk, Civil Engineer, for his assistance. And last, but not least, to my beloved wife, Daisy, whose moral, spiritual and physical support made all my efforts a genuine source of personal satisfaction.

Illustrative material was furnished by City of Los Angeles Bureau of Engineering, Survey Division; Los Angeles County Engineer's Survey Division — Geodetic Section; California Department of Transportation; U.S. Geological Survey; the U.S. Army, the U.S. Bureau of Land Management, the Southwest Engineering Co. and Williamson & Schmid, Engineers. The Keuffel and Esser Company furnished the illustrations of drafting tools and equipment.

G.H.W.

Contents

XI

Chapter 2 *Plotting Traverses*

Chapter 3 *Relating Maps to Descriptions*

Chapter 4 *Lettering and Lines*

Chapter 5 *Drafting Tools*

Chapter 5 *Drafting Tools* cont'd

Chapter 6 *Production of Topographic Maps*

Chapter 7 *Plan and Profile Drawings*

Chapter 11 *Verticle Curves* *cont'd*

Chapter 12 *Special Type Maps*

Chapter 13 *Geographic Control*

Chapter 13 Geographic Control *cont'd*

Chapter 14 *Computerized Drafting*

Appendix

Index

Chapter 1

Field Notes

Reading Survey Field Notes

A very excellent treatise on the methods and procedures of making legible and understandable field notes has been prepared by a well-qualified land surveyor, the late F. William Pafford;[1] therefore, we will discuss not the way to do it but the application through interpretation of the notes into the appropriate maps and drawings. There are a number of fields to be covered in this operation.

Topography refers to the features on the ground as well as the various slopes. Field notes and maps of this nature are used for planning developments of polygonic areas, large or small, or for the alignment and construction of pipelines, railroads, highways, etc. along strips of land.

Field notes of property boundary surveys may be translated into maps used for either a single ownership parcel of land or a subdivision of land into multiownerships or for the acquisition of strip easements or for the determination of the position of an agreement line, or for a court case presentation.

Information obtained by the survey crew in the field is meticulously noted in a book which is later given to the draftsman to translate and transpose onto sheets of paper, cloth or mylar. This is part of the foundation and framework of the job.

One type of information relates to control surveys which find or establish points and lines to be subsequently used for a specific acquisition of land such as for a right of way. Another source of facts is furnished by the crew making a survey of the boundary lines of property ownership. This in part is a matter of recovering points assumed to have been set and hence determining whether they were or were not set and in case any are missing to establish or reset them. Or, that crew may have the job of establishing all new points and lines for the purpose of a sale and from their field notes a map will be made to show the configuration of the parcel and its area, and its juxtaposition to other parcels.

Still another type of control work is done in geodetic surveying which involves the extension of high order horizontal and vertical precise measurements of distances, angles and differential elevations along with determining astronomic longitudes, latitudes and azimuths. These necessarily also require consideration of the shape of the earth as it affects field observations and hence includes appropriate adjustments and computations applicable to those observations.

In construction work the control line and topographic surveys are done first in order to furnish that information to the designer in the office. An important part of this preliminary work is the recording of tie lines to salient control points and bench marks for easy recovery. Subsequently, the field crew recovers the control points, sets the controls for the construction of the improvements and establishes the finish grades.

Engineering surveying refers to the development of horizontal and vertical control into design data which determines the final location of a

[1]Handbook of Survey Notekeeping, published by John Wiley and Sons, Inc., N.Y.

building site or road location with its attendant elevations, alignment, drainage, etc.

Field notes have various forms and styles appropriate to the particular subject involved. The following are samples of selected subject matter just to give you some acquaintance with a cross section of various kinds of field notes. Besides these, there are many others used for special projects and different purposes, but the objective here is to give you an idea of some of the more common types which you will find in an engineering or survey office and with which you must work.

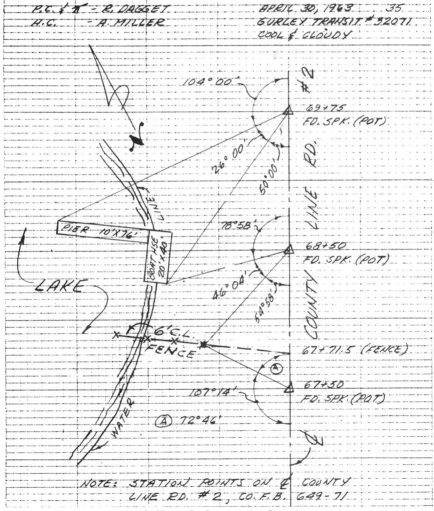

Angle Notes for Intersection Locations

Fig. 1-1

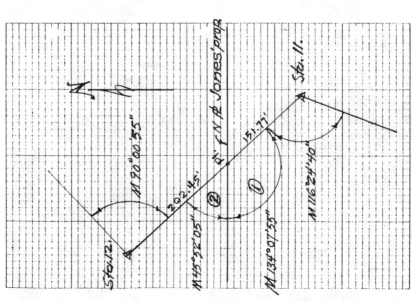

Angle Notes for Traverse with Diagram

Fig. 1-2

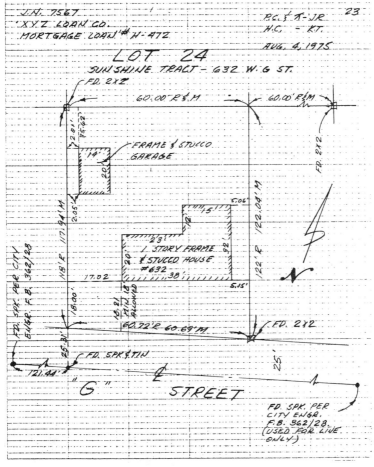

Notes of Mortgage-Loan Survey
Fig. 1-3

EXERCISE 1-1: Draw a page of your own field notes similar to the pattern of Figure 1-3. Use your home site or one nearby. Establish a control line on one side or the other and measure the offsets to the buildings, then measure the setbacks from the walk or curb or street line in front. If you know, or can determine, the size and shape of the lot, then relate your tie distances to the lot lines. Be sure you measure every side of the structure and that the sum of the parts on one side equals the sum of the parts on the opposite side. Finally, draw a map to scale from the notes you made and compare it with the ground conditions.

* * *

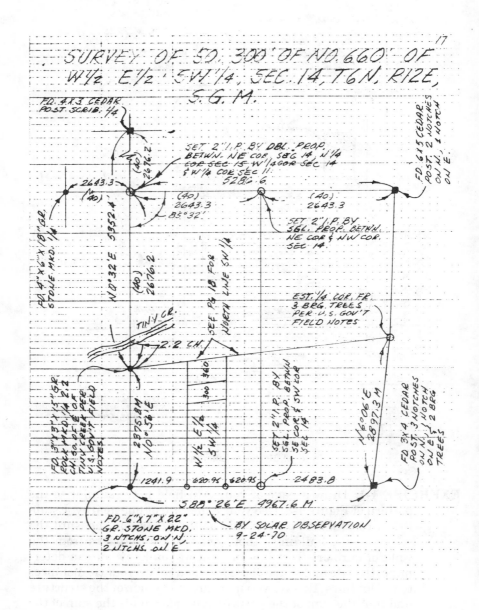

Survey Notes for Boundary Control
of W½ of E½ OF SW¼ Sec 14, T 6 N, R 12 E, S.G.M.
Fig. 1-4a

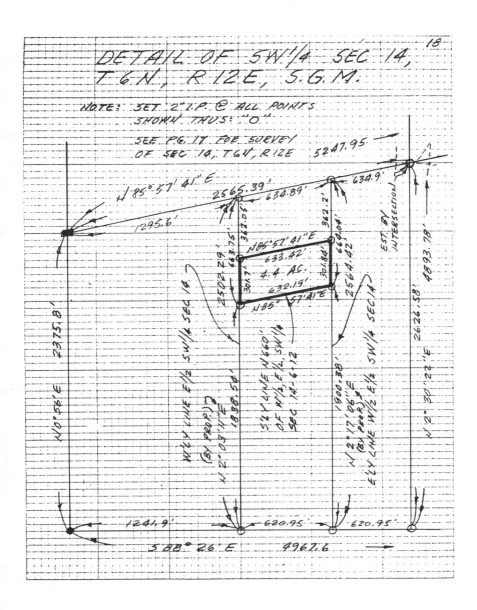

Detail Notes of establishment of S. 300.00 ft.
of N 660.00 ft. W½ E½ SW¼ Sec. 14, T6N, R12E, S.G.M.
Fig. 1-4b

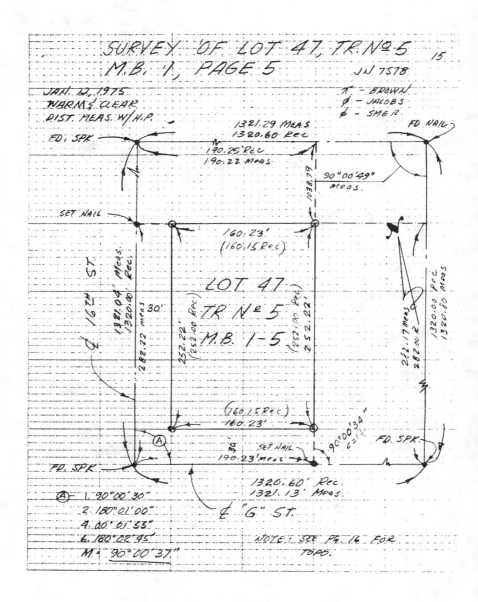

Boundary Notes for Survey of MacMillan's Parcel
Fig. 1-5

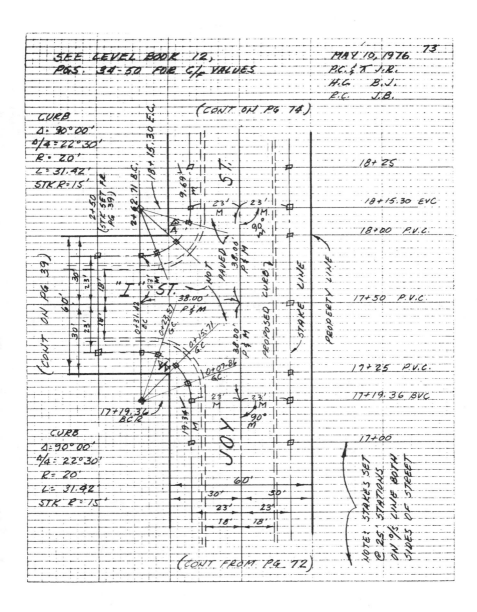

Construction Notes for Curb Stakes
Fig. 1-6

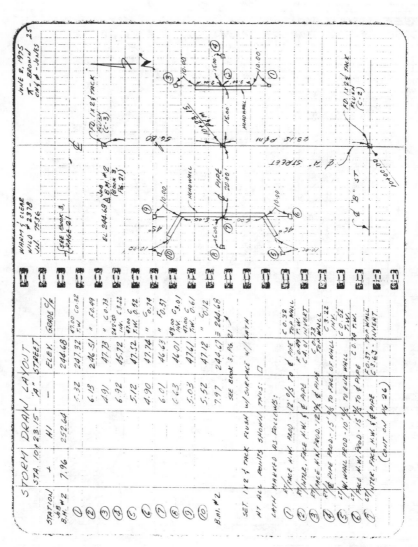

Construction Notes for Storm Drain Under roadway
Fig. 1-7

STATION	+	HI	−	ELEV	GRADE C/E
			(CONT ON PG 4)		
4+00			4.21	811.15	06.34 C 4.81
+75			4.54	810.82	06.19 C 4.63
+50			4.61	10.75	06.04 C 4.71
+25			4.87	10.49	05.89 C 4.60
3+00			5.18	10.18	05.74 C 4.44
+75			5.42	9.94	05.59 C 4.35
+50 INLET N. 2+55 MH			5.64	9.72	05.48 C 4.24
RIM 2+55					810.89 F.S.
OUTLET S. 2+55 MH			5.64	9.72	F.1175 S. 05.38 C 4.34
+25			5.81	9.55	805.22 C 4.33
2+00	6.28	815.36	6.15	809.21	805.07 C 4.14
TP₅ +75			3.76	809.08	04.92 C 4.16
+50			4.09	08.75	04.77 C 3.98
+25			4.30	08.54	08.62 C 3.92
1+00			4.61	08.23	04.47 C 3.76
+75			5.01	07.83	04.32 C 3.51
+50			5.33	07.51	04.17 C 3.34
+25			5.41	807.43	804.02 C 3.61
INLET N. 0+00			5.37	807.47	805.87 C 3.60
B.M.#3 0+00	5.24	812.84		807.60	

SEWER STAKES - APPLE DRIVE BETWEEN FILLMORE & BIRCH

NOTE: SEE SEWER PLANS FOR B.M. USED
SET 112 W/TACK & GUARD STK. MRKD. 10'O/S-E-E STA.

MAY 3, 1975   3

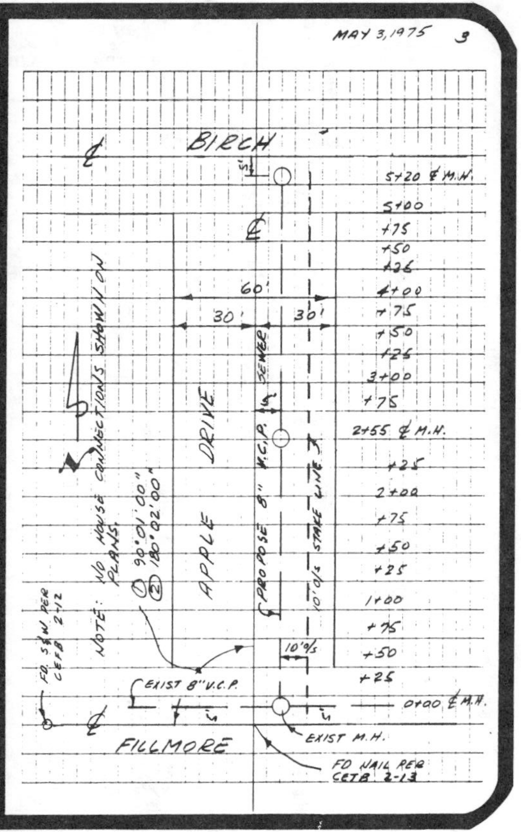

Construction Notes for Sewer Line
Fig. 1-8

Note in Figure 1-9 that the rod reading is on top of the line and over the distance (right or left) from the center line control. In this case you must reduce the reading to give you the correct level for mapping.

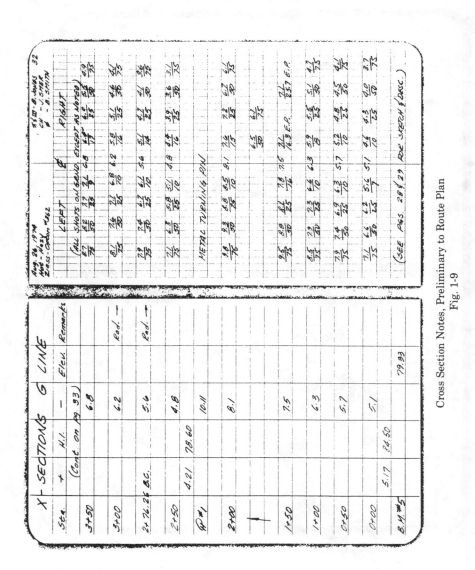

Cross Section Notes, Preliminary to Route Plan

Fig. 1-9

Figure 1-10 shows a common form from which contour lines can be developed on the map.

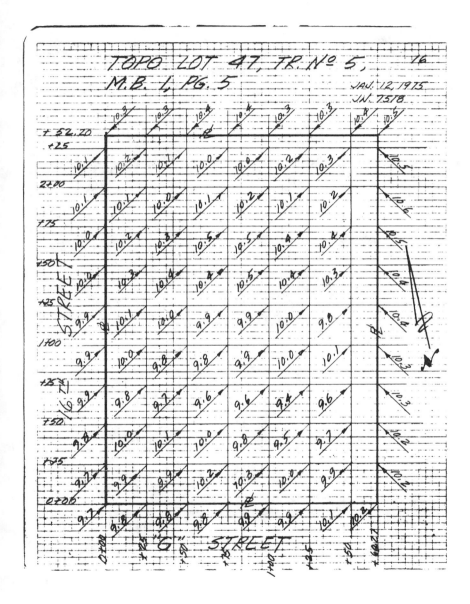

Cross Section Notes on 25-foot Grid. A 10-meter grid
would probably be used in the metric form.
Fig. 1-10

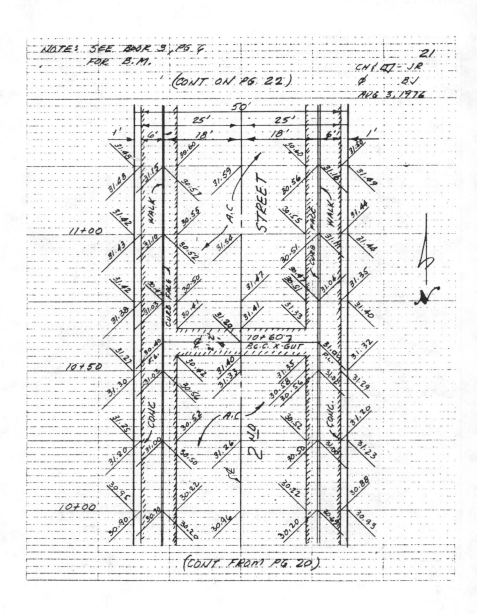

Diagrammatic Cross Section Notes
Fig. 1-11

Eng. 3.882 (Rev. 5-74)

Date 12-4-74   Km Grid 258-282   F.B. 40006   Pg. 16

Return to
BUREAU OF ENGINEERING
SURVEY DIVISION
200 N. Main St. — Los Angeles 90012

NOTE: Additions or changes must be initialed and dated.

Job Number 4-9999
Sheet 4 of 5

Limits Earle St.: Dona Ave. to Mesloh Rd.
B.M. Adjustment Year 1963 Adj.:—B.M. Levels

Station	Set/+	H.I.	Read/-	Elev.	Reference	Stakes	
B.M.	4.94	237.99		233.05	06-0130	S.S.M. on c/l inter. Boundary Rd.	
						₵ Earle St.(W)	
6+45			4.85	233.14		Pav.	37' Lt. of Rt. R/W
6+46.9			5.04	232.95		"	43.7'
6+47.1			4.61	233.38		Conc. Drv.	50'
"			4.04	233.95		"	65'
6+50			4.0	234.0		Grd.	10' Rt. of Rt. R/W
"			3.57	234.42		Conc. Box	1' Lt. of Rt. R/W
"			5.3	232.7		Grd.	1'
"			4.58	233.41		Pav.	13'
"			4.39	233.60		"	25'
"			5.01	232.98		"	37'
"			4.42	235.57		Conc. Drv.	50'
"			4.10	233.89		"	65'
6+55.80			4.59	233.40		"	50'
"			4.19	233.80		"	65'
6+60.2			5.11	232.88		4" Weep Hole	43.6' "
6+73			4.4	233.6		Grd.	10' Rt. of Rt. R/W

Micro-film   516 M   Index

Level Notes for Street Improvement
Fig. 1-12

Eng. 3.582 (Rev. 5-74)

Date 12-4-74 — Km Grid 258-282 — F.B. 40006 — Pg. 16

Return to
BUREAU OF ENGINEERING
SURVEY DIVISION
200 N. Main St. — Los Angeles 90012

NOTE. Additions or changes must be initialed, and dated.

Limits Earle St.: Dona Ave. to Masloh Rd.
B.M. Adjustment Year 1963 Adj. — B.M. Levels
Job Number 4-9999
Sheet 4 of 5

Station	+(B.S.)	H.I.(m)	Rod(m)	El.(m)	Reference	Stakes
B.M.	1.506	72.539		71.034	06-01320	SSM on c/l inter. Boundary Rd. & Earle St. (W)
2+08.48 E			1.478	71.061		Pav. 11.28m Lt. of Rt. R.
2+09.05 E			1.536	71.003		" 13.32m "
2+09.12 E			1.405	71.134		Conc. Drv. 15.24m "
" E			1.231	71.308		" 19.81m "
2+10 W			1.22	71.32		Grd. 3.05m Rt. of Rt. R.
" W			1.088	71.451		Conc. Box 0.30m Lt. of Rt. R.
" W			1.62	70.92		Grd. 0.30m "
" W			1.396	71.143		Pav. 3.96m "
" c/L			1.338	71.201		" 7.62m "
" E			1.527	71.012		" 11.28m "
" E			1.347	71.192		Conc. Drv. 15.24m "
2+11.77 E			1.249	71.290		" 19.81m "
" E			1.399	71.140		" 15.24m "
" E			1.277	71.262		" 19.81m "
2+13.11 E			1.557	70.982		0.10m Weep Hole 13.29
2+17.01 W			1.34	71.20		Grd. 3.05 Rt. of Rt. R.

Micro-film 516 M   Index

Same job shown in Figure 1-12 done in metric.
Fig. 1-13

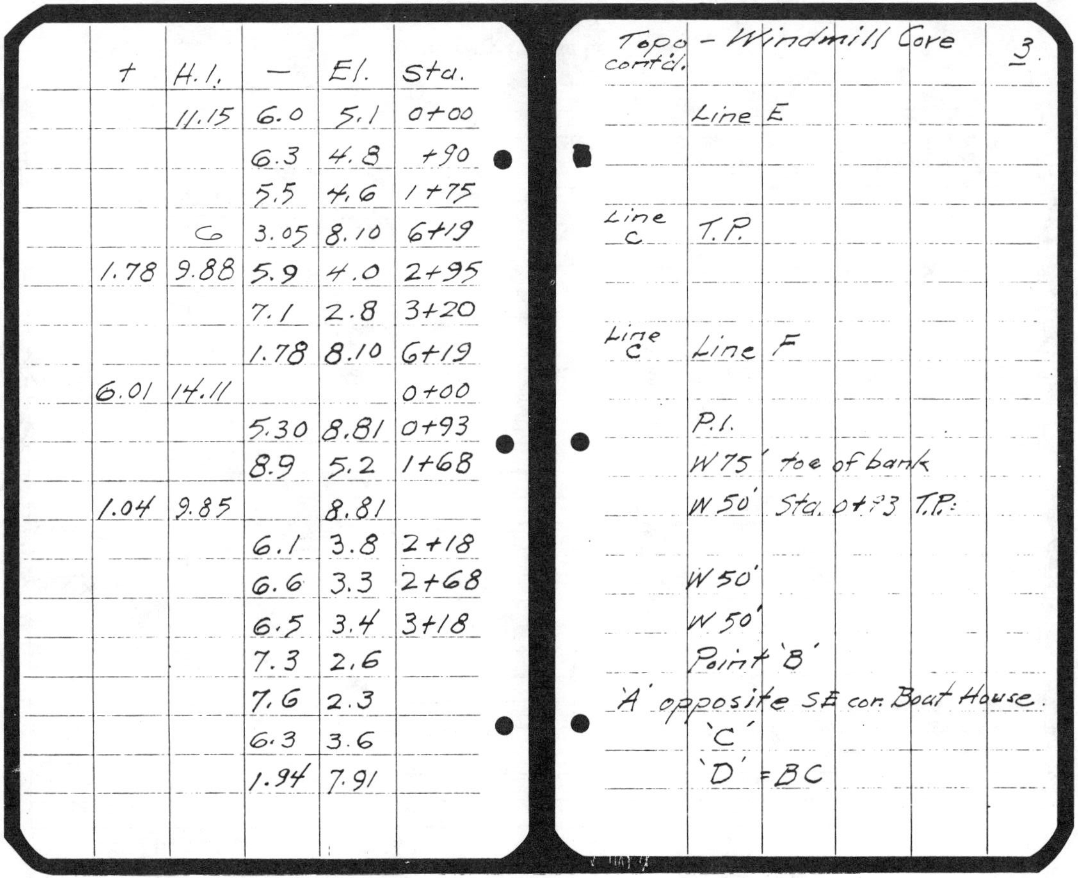

Topo - Windmill Cove cont'd.   3

+	H.I.	−	El.	Sta.	Notes
	11.15	6.0	5.1	0+00	
		6.3	4.8	+90	
		5.5	4.6	1+75	Line E
C		3.05	8.10	6+19	
1.78	9.88	5.9	4.0	2+95	Line C T.P.
		7.1	2.8	3+20	
		1.78	8.10	6+19	Line C Line F
6.01	14.11	5.30	8.81	0+93	P.I.
		8.9	5.2	1+68	W75' toe of bank
1.04	9.85	8.81			W 50' Sta. 0+73 T.P.
		6.1	3.8	2+18	W50'
		6.6	3.3	2+68	W50'
		6.5	3.4	3+18	Point 'B'
		7.3	2.6		'A' opposite SE cor Boat House
		7.6	2.3		'C'
		6.3	3.6		'D' = BC
		1.94	7.91		

Level Notes (single wire) Windmill Cove
Fig. 1-14

Eng. 3.882 (Rev. 5-74)

Date 12-2-74　　Km Grid 259-277　　F.B. 44444　　Pg. 333

Return to
BUREAU OF ENGINEERING
SURVEY DIVISION
200 N. Main St — Los Angeles 90012

NOTE: Additions or changes must be initialed, and dated.

Job Number 6-9999
Sheet 15 of 17

Limits: Crenshaw Blvd., Bronson Ave.
B.M. Adjustment Year Basis: 1963 Precise

Station	Set/+	HI	Read/-	Elev.	Reference	Stakes
B.M.	5.119	164.705		159.586	12-15190	Spk N. curb Olympic Blvd., 15.5'W of B.C. ret. NW cor. Crenshaw Blvd.
T.P.	3.630	163.535	4.800	159.905		
T.P.	4.960	162.320	6.175	157.360		
B.M.&T.P.	5.845	162.675	5.490	156.830	No Ref.	Fd Spk W.curb Crenshaw Bl, 12'S/o Country Club Dr. (S. end C.B.)
B.M.&T.P.	6.140	163.075	5.740	156.935	No Ref.	Fd Spk E.curb Crenshaw Bl, 11.5'S/o Country Club Dr. (S. end C.B.)
T.P.	4.420	163.510	3.985	159.090		
B.M.&T.P.	5.735	163.605	5.640	157.870	No Ref.	Fd spk W.curb Bronson Ave, 8'S/o Country Club Dr.
B.M.&T.P.	6.060	164.265	5.400	158.205	New	Set Spk E. curb Bronson Ave, 2.5'N/o N to prod Country Club Dr.
T.P.	6.100	164.910	5.455	158.810		
B.M.&T.P.	5.950	166.120	4.740	160.170	New	Set Spk N.curb Olympic Bl, 7.5'E/o E. to prod Bronson Ave.
T.P.	5.425	166.550	4.995	161.125		
B.M. Chk.			5.080	161.470	161.444 (12-15177)	Wire Spk W.curb Norton Ave, 23.5' N/o B.C. N/o Olympic Bl. (like old C.B)

Micro-film ① 517M ② ③ ④ ⑤ ⑥ ⑦　Index

Single Wire Precise Level Run
Fig. 1-15

Eng. 3.882 (Rev. 5-74)   Date 12-2-74   Km Grid 259-277   F.B. 44444   Pg. 333

NOTE: Additions or changes must be initialed, and dated

Return to
BUREAU OF ENGINEERING
SURVEY DIVISION
200 N. Main St. — Los Angeles 90012

Limits: Crenshaw Blvd, Bronson Ave.
B.M. Adjustment Year Basis: 1963 Precise

Job Number 6-9999
Sheet 15 of 17

Station	Sight + (m)	HI (m)	Read - (m)	Elev (m)	Reference	Stakes
① B.M.	1.560	50.202		48.642	12-15190	Spk. N. curb Olympic Blvd., 4.7m W of B.C. ret. NW cor. Crenshaw Blvd.
T.P.	1.106	49.846	1.463	48.739		
T.P.	1.512	49.475	1.882	47.963		
② B.M.§T.P.	1.782	49.583	1.673	47.802	No Ref.	Fd. spk. W. curb Crenshaw Bl.,3.7m S/o Country Club Dr. (S. end C.B.)
③ B.M.§T.P.	1.871	49.705	1.750	47.834	No Ref.	Fd. spk. E. curb Crenshaw Bl.,3.5m S/o Country Club Dr. (S. end C.B.)
T.P.	1.347	49.838	1.215	48.491		
④ B.M.§T.P.	1.748	49.867	1.719	48.119	No Ref.	Fd. spk. W. curb Bronson Ave.,2.4m S/o Country Club Dr.
⑤ B.M.§T.P.	1.847	50.068	1.646	48.221	New	Set spk. E. curb Bronson Ave.,0.8m N/o R prod. Country Club Dr.
T.P.	1.859	50.265	1.663	48.405		
⑥ B.M.§T.P.	1.814	50.633	1.445	48.820	New	Set spk. N. curb Olympic Bl., 2.3m E/o E. R prod. Bronson Ave.
T.P.	1.654	50.764	1.522	49.111		
⑦ B.M.Chk.			1.548	49.216	49.208 (12-15177)	Wire spk. W. curb Norton Ave. 7.2m N/o B.C. N/o Olympic Bl. (N.end C.B.)

Micro-film 517 M   Index

Same Precise Level Run as in Figure 1-16 run and recorded in metric.
Fig. 1-16

Figure 1-17 illustrates the 3-wire (horizontal hairs in the telescope) method for precise control leveling.

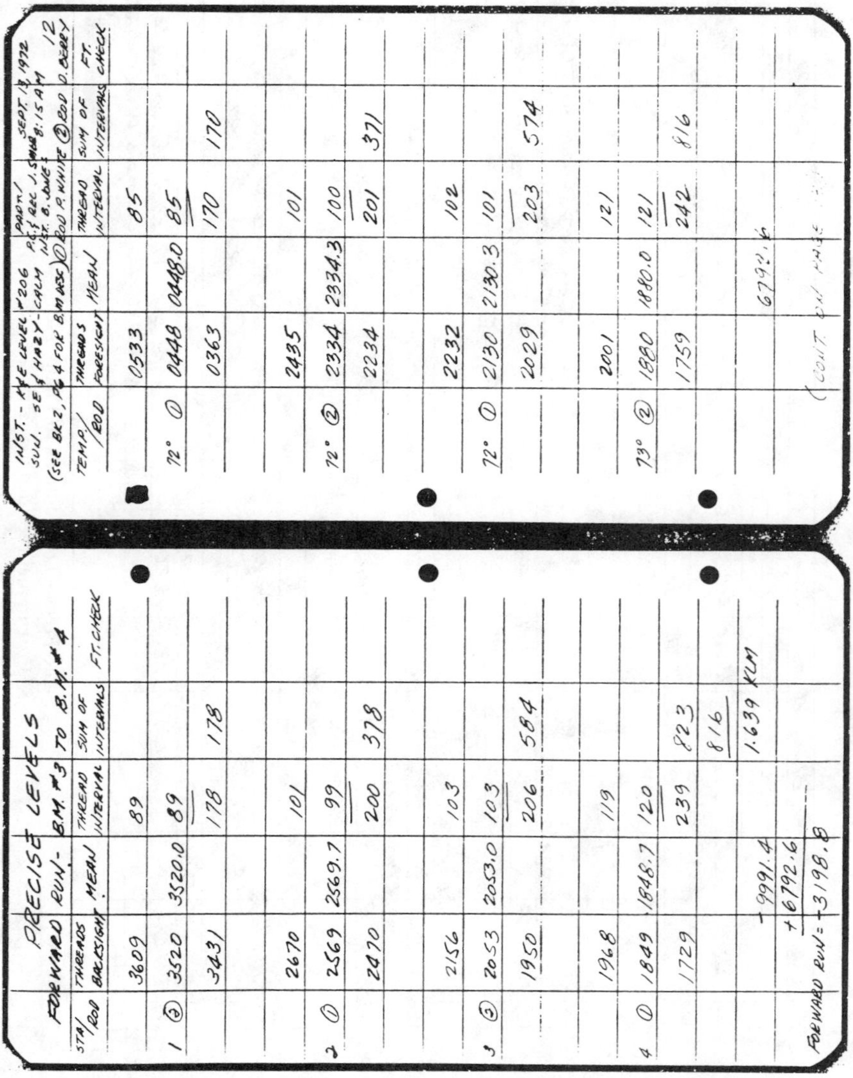

INST. - K&E LEVEL #206 PARTY: ... SEPT 18 1972
SUJ. SEE HAZY-CALM ... 8:15 AM
(SEE BK 2, P. 4 FOR BM...)

Check Run (Foresight page)

TEMP/ROD	THREADS FORESIGHT	THREAD MEAN	THREAD INTERVAL	SUM OF INTERVALS	FT. CHECK
	0533		85		
72° ①	0448	0448.0	85		
	0363		170	170	
	2435		101		
72° ②	2334	2334.3	100		
	2234		201	371	
	2232		102		
72° ①	2130	2130.3	101		
	2029		203	574	
	2001		121		
73° ②	1880	1880.0	121		
	1759		242	816	
		6792.6			

(CONT ON NEXT PAGE)

PRECISE LEVELS
FORWARD RUN - BM #3 TO B.M #4

STA/ROD	THREADS BACKSIGHT	THREAD MEAN	THREAD INTERVAL	SUM OF INTERVALS	FT. CHECK
	3609		89		
1 ①	3320	3320.0	89		
	3431		178	178	
	2670		101		
2 ①	2569	2569.7	99		
	2470		200	378	
	2156		103		
3 ②	2053	2053.0	103		
	1950		206	584	
	1968		119		
4 ①	1849	1848.7	120		
	1729		239	823	
				816	
				1.639 KLM	

-9991.4
+6792.6
FORWARD RUN = -3198.8

Precise Level Notes (3-wire) Check Run
Fig. 1-17

PROFILE - FROM N.E. COR. PRKG.
LOT TO EXIST. 24"Ø RCP DRAIN

Sta.	+	H.I.	−	Elevs. TOP&BMs	Profile
B.M. #36	3.71	440.92		437.21	
Ⓟ	5.10	440.40	5.62	435.30	
0+00			5.20		435.20
0+50			5.3		435.1
1+00			5.5		434.9
1+30			5.7		434.7
1+33			7.3		433.1
1+36			7.4		433.0
1+40			5.6		434.8
1+50			5.9		434.5
2+00			6.1		434.3
2+41			9.5		430.9
2+41			6.4		434.0
2+50			6.6		433.8
Ⓟ2+65.50	4.31	438.34	6.37	434.03	433.8
3+00			5.0		433.3
3+25			5.2		433.1
3+25			5.7		432.6
Ⓟ3+47.3			5.31	433.03	
Σ 13.12			17.30		
(SEE PG. 36)					

(NOTE: SEE PGS. 30-34 FOR HORIZONTAL
LOCATION & STAKE OUT OF ROUTE)   35

MAY 12, 1972
WARM & CLEAR

R.E. SPK IN POWER POLE   T - B. JONES
#6Y6783 E. SEE PG. 30   ∮ - B. SMITH
METHL TURNING PIN   CH∮CO - J. SMER

SPK∮TH @ HIGH POINT N PVMT. WILD #7265
N.E. COR. PARKING LOT.
GROUND
"

" TOP OF DITCH   + 13.12
" BOT " " "   − 17.30
" " " " "   − 4.18
" TOP " "   437.21
                      433.03 ✓

TOP OF 4"Ø C.I. PIPE
GROUND @ PIPE
GROUND

TOP OF 2½ STAKE @ ⅓ PT, GROUND
0.2' BELOW TOP OF STAKE
GROUND
T.C.
F.L.
TOP OF SPK∮TH @ ∮ - SEE PG. 34

Profile — Single Line Run
Fig. 1-18

Note in Figure 1-19 that the actual levels are to be reduced in the office from the rod readings recorded in the field.

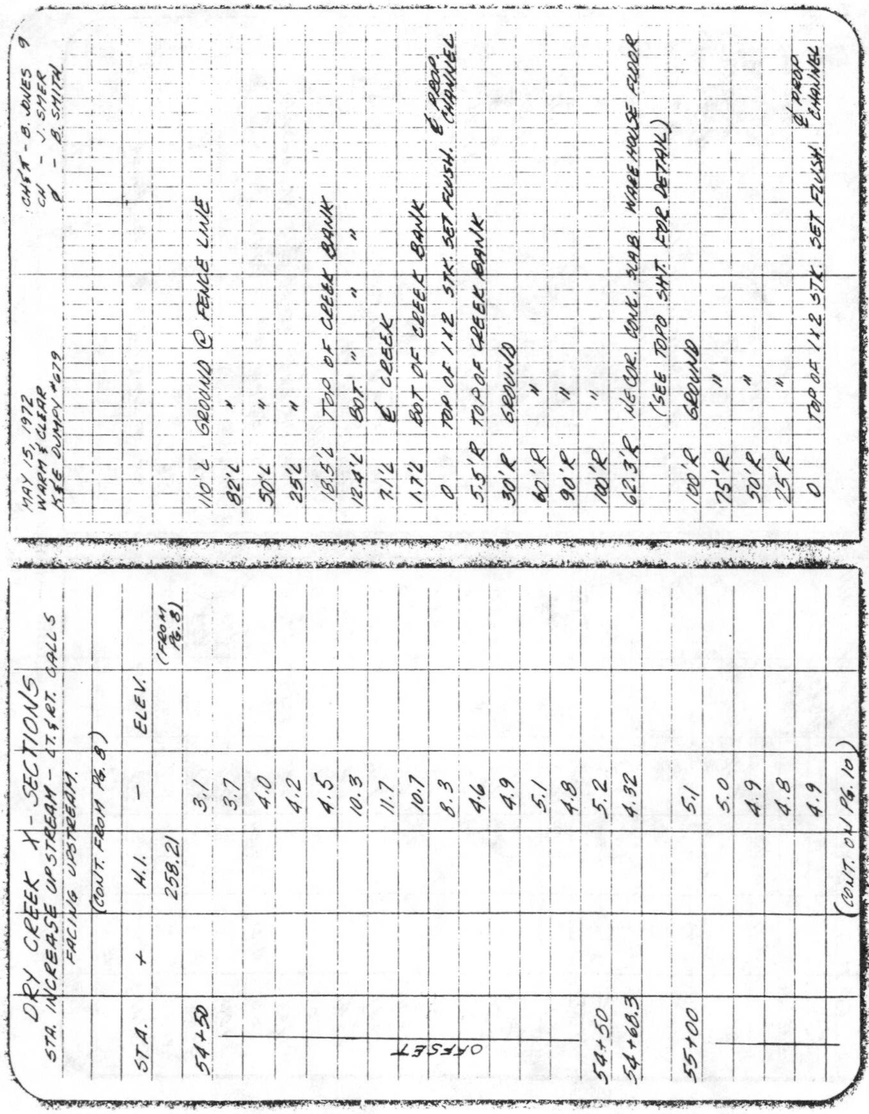

DRY CREEK X-SECTIONS
STA. INCREASE UPSTREAM — LT. & RT. CALLS
FACING UPSTREAM.

MAY 15, 1972
WARM & CLEAR
K&E DUMPY #279

CH&F — B. JONES
CH — J. SMER
∅ — B. SMITH

(CONT. FROM PG. 8)

STA.	+	H.I.	−	ELEV (FROM PG. 8)	CALLS
54+50		258.21			
			3.1		110'L GROUND @ FENCE LINE
			3.7		82'L "
			4.0		50'L "
			4.2		25'L "
			4.5		18.5'L TOP OF CREEK BANK
			10.3		12.4'L BOT " "
			11.7		7'L ₵ CREEK
			10.7		1.7'L BOT OF CREEK BANK
			8.3		0 TOP OF 1×2 STK. SET FLUSH. ₵ PROP. (CHANNEL)
			4.6		5.5'R TOP OF CREEK BANK
			4.9		30'R GROUND
			5.1		60'R "
			4.8		90'R "
			5.2		100'R "
54+68.3			4.92		62.3'R N.G COR. CONC. SLAB. WARE HOUSE FLOOR (SEE TOPO SHT. FOR DETAIL)
55+00			5.1		100'R GROUND
			5.0		25'R "
			4.9		20'R "
			4.8		25'R "
			4.9		0 TOP OF 1×2 STK. SET FLUSH ₵ PROP CHANNEL

(CONT. ON PG. 10)

Profiles Notes — With Cross Sectioning
Fig. 1-19

Preliminary reconnaissance by stadia shown in Figure 1-20 is not intended to be an accurate control survey.

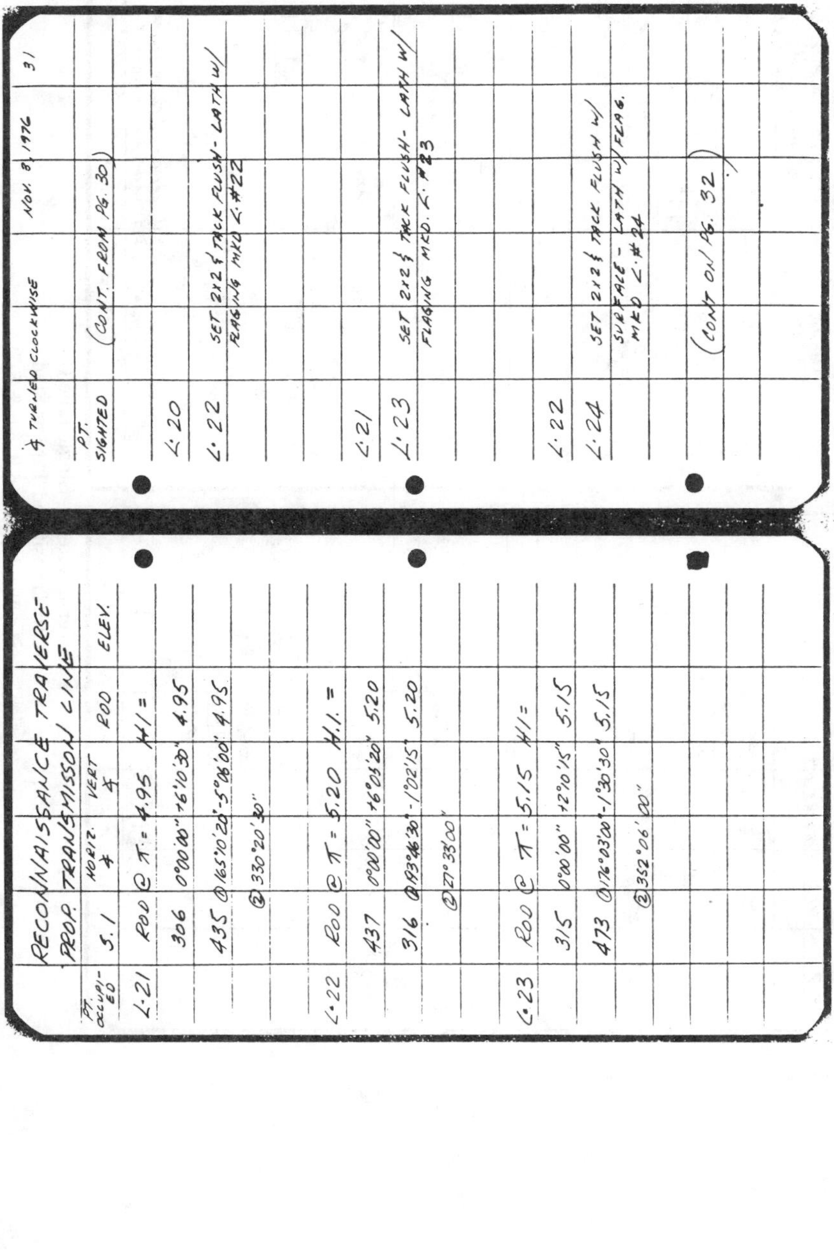

RECONNAISSANCE TRAVERSE
PROP. TRANSMISSION LINE

PT. OCCUPIED	HORIZ. ⊄	VERT. ⊄	ROD	ELEV.
L.21	R.O.O @ T.= 4.95 H.I. =			
306	0°00'00"	-6°10'30"	4.95	4.95
435	@165°10'20"	-5°08'00"	4.95	
	@330°20'30"			
L.22	R.O.O @ T.= 5.20 H.I. =			
437	0°00'00"	+6°03'20"	5.20	
3/6	@83°46'30"	-1°02'15"	5.20	
	@27°33'00"			
L.23	R.O.O @ T.= 5.15 H.I. =			
3/5	0°00'00"	+2°40'15"	5.15	
473	@176°23'00"	-1°30'30"	5.15	
	@352°06'00"			

⊄ TURNED CLOCKWISE   NOV. 8, 1976   31

PT. SIGHTED	
	(CONT. FROM PG. 30)
L.20	
L.22	SET 2x2 & TACK FLUSH- LATH W/ FLAGING MKD L.#22
L.21	
L.23	SET 2x2 & TACK FLUSH- LATH W/ FLAGING MKD. L.#23
L.22	
L.24	SET 2x2 & TACK FLUSH W/ SURFACE- LATH W/ FLAG. MKD L.#24
	(CONT. ON PG. 32)

Traverse Notes — Reconnaissance
by Stadia
Fig. 1-20

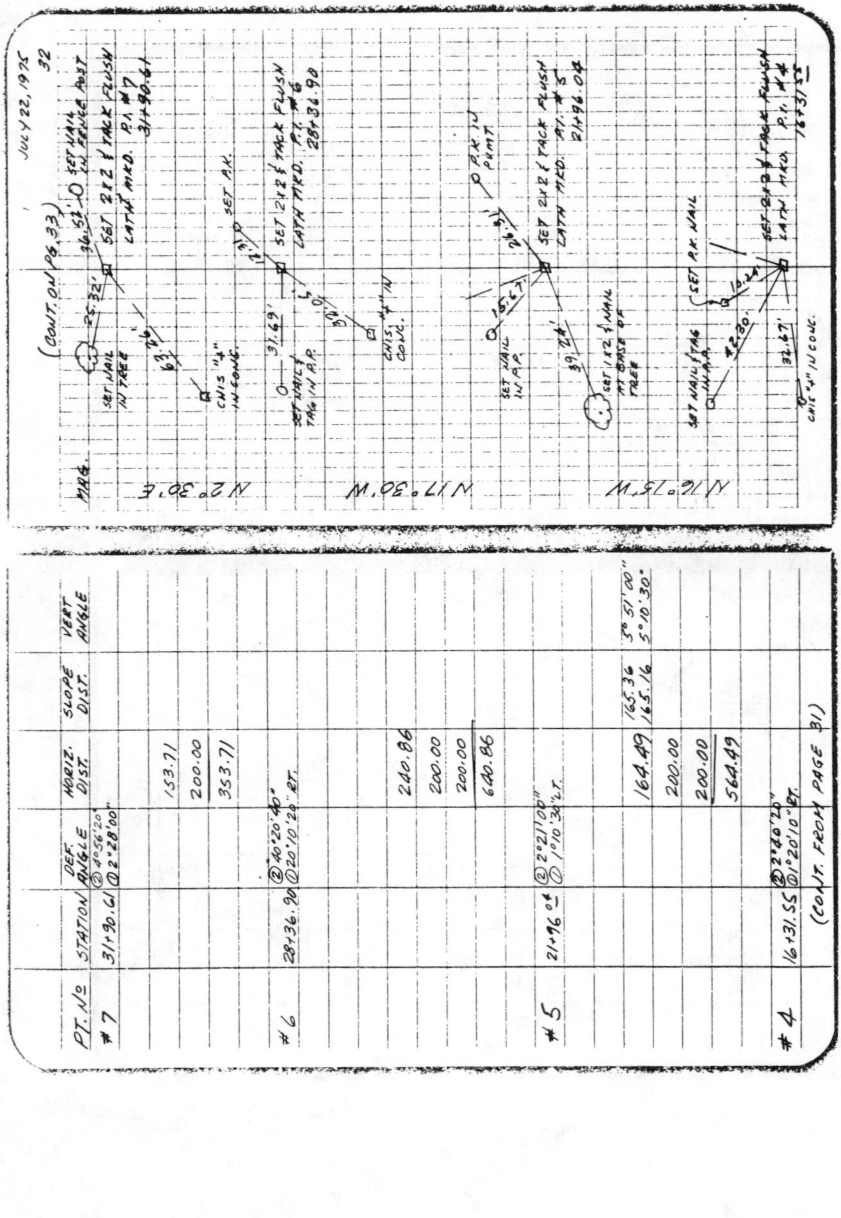

JULY 22, 1975   32

(CONT. ON PG. 33)

MAG.   N2°30'E   N17°30'W   N17°30'W   N16°15'W

PT. N°	STATION	DEF. ANGLE	HORIZ. DIST.	SLOPE DIST.	VERT. ANGLE
#7	31+90.61	@4°56'20" @2°28'00"			
			153.71		
			200.00		
			353.71		
#6	28+36.90	@40°20'40" @20°10'20" RT			
			240.86		
			200.00		
			200.00		
			640.86		
#5	21+96.04	@2°21'00" @1°10'30" LT			
			164.49	165.36	3°51'00"
			200.00	165.16	5°10'30"
			200.00		
			564.49		
#4	16+31.55	@2°40'20" @1°20'10" RT			
	(CONT. FROM PAGE 31)				

Traverse Notes — Single Line Open
With Transit and Tape or Stadia
(See also Figures 6-3, 6-6 & 6-7)
Fig. 1-21

A precise traverse set with transit and tape, whether it be single line open or closed, as shown in Figures 1-21 and 1-22, is for the purpose of controlling the information obtained around each station on the line.

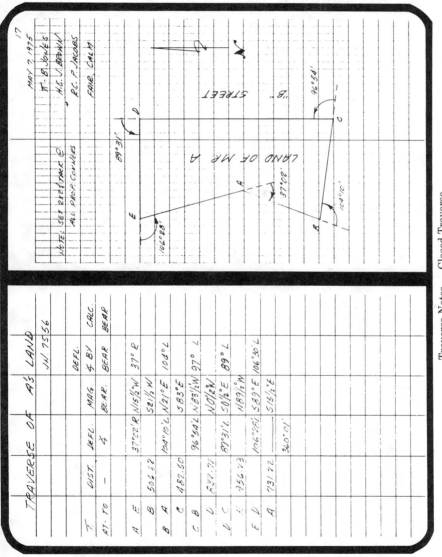

Traverse Notes — Closed Traverse
with Transit and Tape
Fig. 1-22

Sometimes a precise traverse is run for establishing control for a realignment. When the new line is calculated in the office, the field

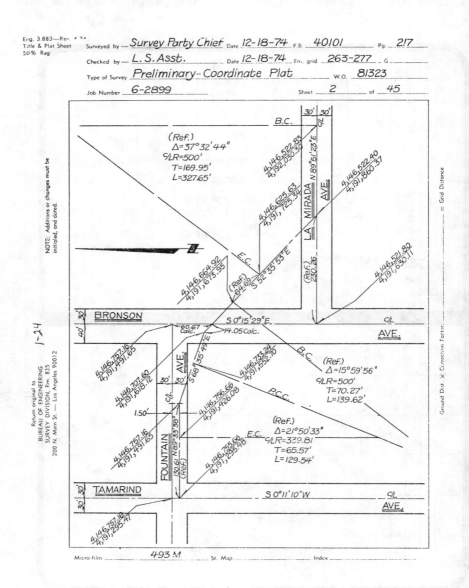

Pre-calculated Field Notes for a realignment job.
Fig. 1-23

notes are also prepared in the office by which the field crew will set and control the new work. Figures 1-23 and 1-24 show this type of field notes.

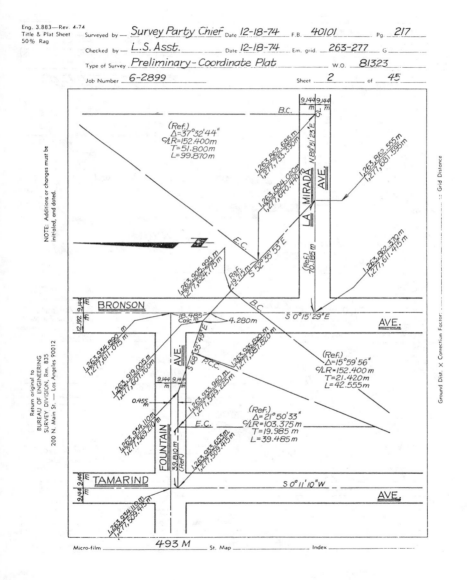

Same job as shown in Figure 1-23 done in metric.
Fig. 1-24

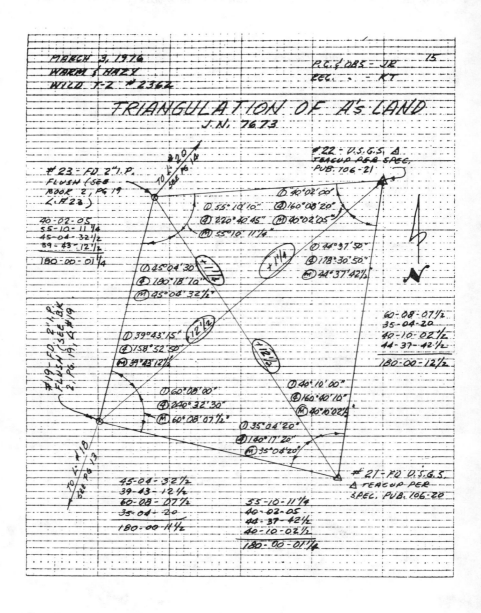

Triangulation Notes
Fig. 1-25

The stadia notes shown in Figure 1-26 are typical for the location of topographical features surrounding a station established on a control traverse.

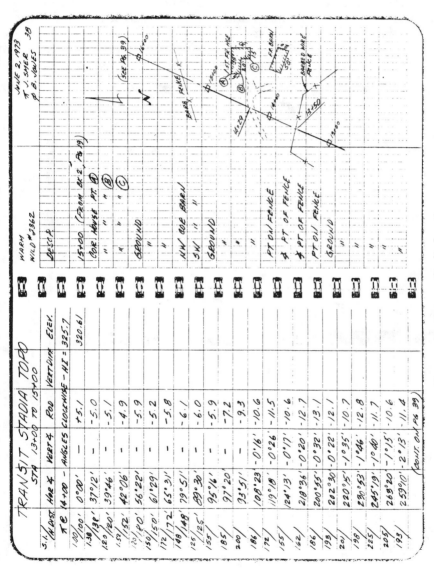

Topographic Notes — Stadia Locations
Fig. 1-26

Hydrographic notes are slightly more involved in that the boat from which the soundings are taken must be positioned by angular or distance intersection and the soundings matched by time position.

By one method, the boat follows a station line (1 + 00, etc.) with signals from a transit on shore at that station to keep it on line while another transit measures an azimuthal angle from a control point to the boat at fixed time intervals. See Figure 1-27.

In producing the hydrographic map, the boat positions are established first, then the soundings are matched to them.

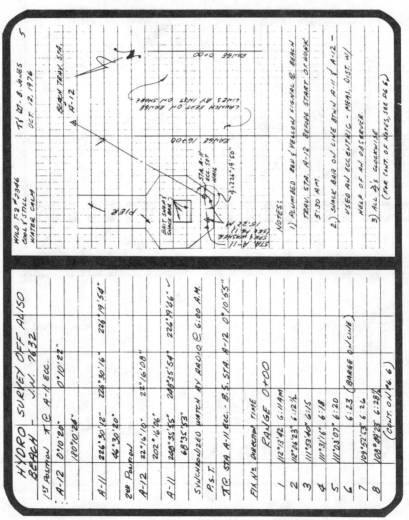

Hydrographic Notes — Boat Fixing by Transit Intersection
Fig. 1-27

The other method requires all operations performed on the boat. Two stations, but preferably three in case one is obstructed from view, and range marks set on lines of even stationing are established on shore. Two operators at the same position on the boat measure angles with their sextants from the range line to their respective land target stations at specified time intervals. See Figure 1-28.

In both cases, soundings for the depth of water are taken at the same timing as the angular observations.

COOL & HIGH CLOUDS LIGHT BREEZE — CALM SEA - 2' GRD, SWELLS FROM SOUTH 20 — JAN. 12, 1975

PARTY & EQUIPMENT OPERATING FROM 32' MOTOR LAUNCH "JOY"

P.C. & BOAT PLOT - B. JONES
SOUND GEAR - J. SMER
"A" SEXTANT & ☐ - G. WHITE
"B" - H. BROWN
HELMSMAN - CAPT. T.T. WIND
"A" SEXTANT #12346 - "B" SEXTANT #0347
CHRONOMETER #2386

BENDIX ECHO SOUNDER # B-382365 - L.M. WITH TRANSDUCER HEAD MOUNTED MIDSHIP ON STARBOARD SIDE & 2.5' BELOW WATER LINE.

SEE J.N. 7522 (BOOK 3, PG. 15) FOR BEACH TRAVERSE AND RANGE LINE LOCATIONS & SIGNAL COLORS.

SEE PAGE 19 FOR SOUND GEAR CALIBRATION.
SEE PAGE 18 FOR STANDARDIZATION OF CHRONOMETER & CHECK OF LAUNCH COMPASS.

STA. 8-12
"A" SEXTANT
"B" SEXTANT
"B" RANGE MARK
TO RANGE MARK
LAUNCH FIX
STA. B-15

HYDRO SURVEY OFF BEACON BAY - J.N. 7523

NOTES:

1.) RANGE LINES RUN ON PRE-DETERMINED BEARING TO RANGE SIGNAL ON SHORE @ LAUNCH SPEED OF APPROX. 5 KNOTS.

2.) BOAT PLOT ON CIRCULAR PLOTTING CHARTS.

3.) SOUNDING GRAPH MARKED & NUMBERED AT EACH "FIX".

4.) SEXTANT ANGLES ALWAYS FROM RANGE MARK TO SIGNAL NOTED.

LAUNCH ON RANGE 10+00
BEARING N 51° 30' E
B-12   B-15

FIX No.	"A" SEXTANT	"B" SEXTANT	TIME
1	42°11'	49°53'	7:02 AM
2	42°36'	50°20'	7:05
3	42°43'	51°08'	7:08½
4	43°07'	51°53'	7:11
5	43°29'	53°08'	7:14
6	43°57'	54°31'	7:17
7	44°26'	55°49'	7:20 AM

(CONT. ON PG. 21)

Hydrographic Notes — Boat Fixing by Sextants
Fig. 1-28

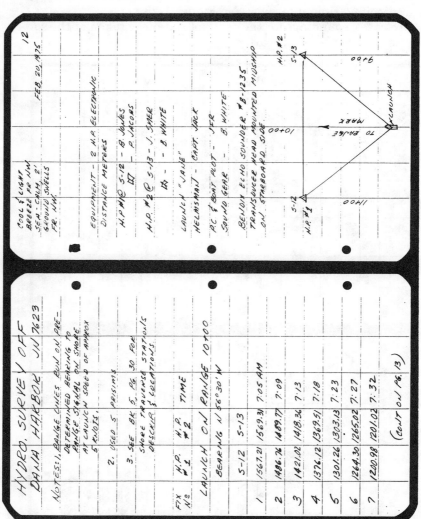

Hydrographic Notes — Boat Fixing by
Electronic Distance Measurement
Fig. 1.29

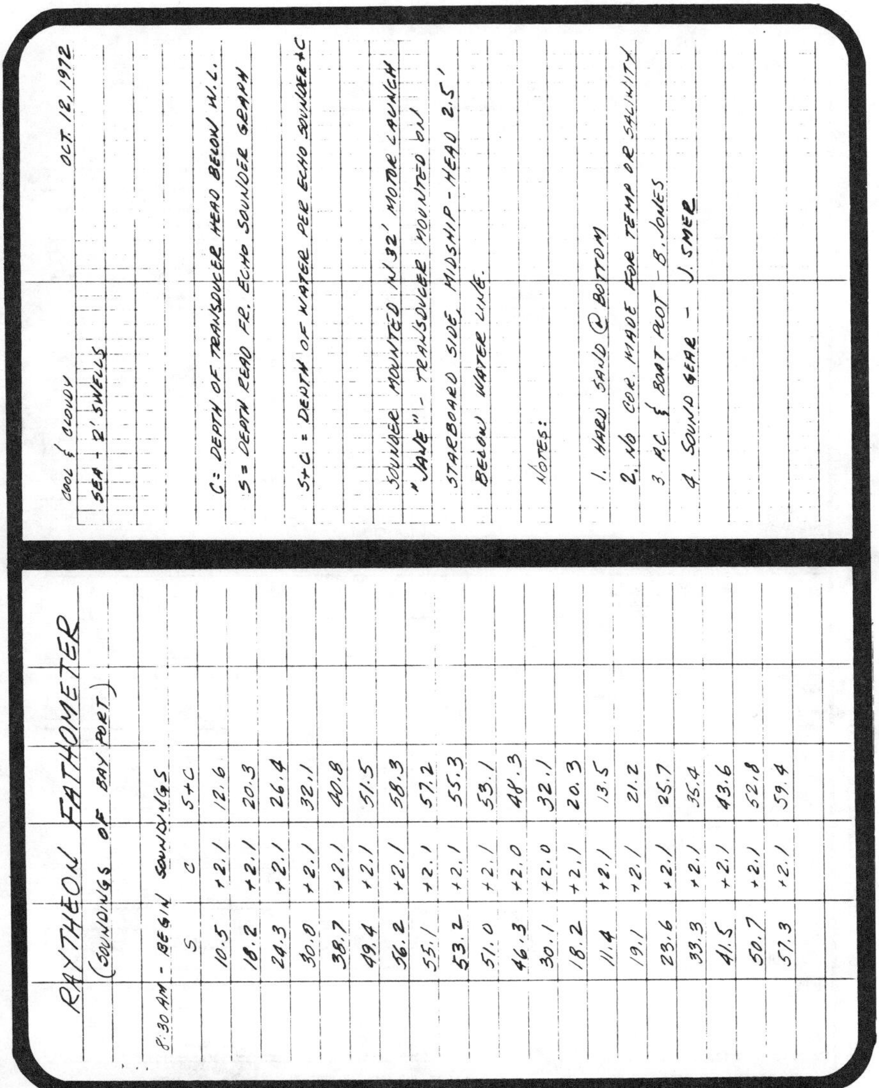

RAYTHEON FATHOMETER

(SOUNDINGS OF BAY PORT)

8:30 AM - BEGIN SOUNDINGS

S	C	S+C
10.5	+2.1	12.6
18.2	+2.1	20.3
24.3	+2.1	26.4
30.0	+2.1	32.1
38.7	+2.1	40.8
49.4	+2.1	51.5
56.2	+2.1	58.3
55.1	+2.1	57.2
53.2	+2.1	55.3
51.0	+2.1	53.1
46.3	+2.0	48.3
30.1	+2.0	32.1
18.2	+2.1	20.3
11.4	+2.1	13.5
19.1	+2.1	21.2
23.6	+2.1	25.7
33.3	+2.1	35.4
41.5	+2.1	43.6
50.7	+2.1	52.8
57.3	+2.1	59.4

COOL & CLOUDY   OCT. 12, 1972

SEA - 2' SWELLS

C = DEPTH OF TRANSDUCER HEAD BELOW W.L.

S = DEPTH READ FR. ECHO SOUNDER GRAPH

S+C = DEPTH OF WATER PER ECHO SOUNDER + C

SOUNDER MOUNTED IN 32' MOTOR LAUNCH "JANE" - TRANSDUCER MOUNTED ON STARBOARD SIDE, MIDSHIP - HEAD 2.5' BELOW WATER LINE.

NOTES:

1. HARD SAND @ BOTTOM

2. NO COR. MADE FOR TEMP OR SALINITY

3. R.C. & BOAT PILOT - B. JONES

4. SOUND GEAR - J. SMEE

Hydrographic Notes — Soundings
Fig. 1-30

The form shown below illustrates the type of information required in keeping field notes when using an electronic measuring instrument such as the Rangemaster.

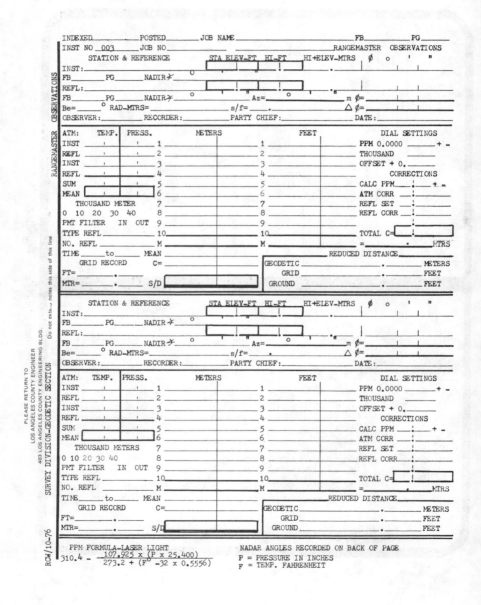

Fig. 1-31

Annotated Field Notes of a Topographic Survey.

The following eight pages are copies of a set of field notes from a survey of John Sims' Lot 15 in the BIRCH WOODS Tract. The annotations will help you to interpret the notes and enable you to more easily follow the footsteps of the surveyor. This will also assist you in understanding other field notes.

First look at the sketch on the left side of page 2 and study the relationships of the lot, Birch Tree Drive and the private driveway on the north side. Look at the dimensions and angles of the lot and realize the shape and size of it.

Next notice the stations marked A, B, C and D at the Lot corners which the surveyor refers to in his notes. Review the level circuit which he ran on these stations to establish controls.

Note that the elevations and topography along Birch Tree Drive and the private driveway are recorded in cross-section form while the interior information is recorded in stadia form.

ANNOTATED

FIELD NOTES

JOB No. 1234
TOPO SURVEY
JOHN SIMS PROP.
LOT 15
BEECH WOODS
TRACT

C. O. P. — JONES
Ⳁ — SMITH
ROD — SMER

COOL AND CLEAR
62° = −0.004/100'
TEMP COR. ON TAPE

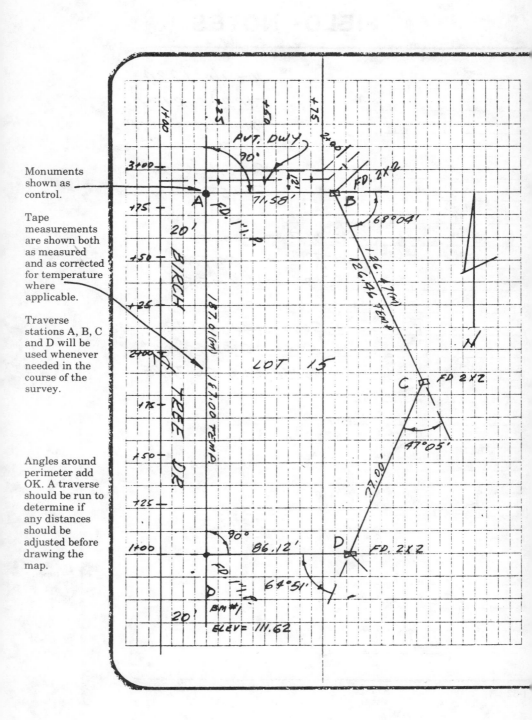

Monuments shown as control.

Tape measurements are shown both as measured and as corrected for temperature where applicable.

Traverse stations A, B, C and D will be used whenever needed in the course of the survey.

Angles around perimeter add OK. A traverse should be run to determine if any distances should be adjusted before drawing the map.

					2
	LEVEL CIRCUIT				
	+	H.I.	−	ELEV	
				111.62	BM#1 SEE BK2, PG3
	11.13	122.75			
GRD. @ 1" I.P. "A"			4.47	118.28	
GRD. @ 2x2 "B"		CHK OK	1.70	121.05	
2x2 @ "C"			4.82	117.93	
2x2 @ "D"			7.39	115.36	
BOLT ON F.H.			11.14	111.61	B.M.#1

	90°00'	
	90 00	
	68 04	
	47 05	
	64 51	
	360°00'	OK

Levels are run on the circuit first to be sure they close within allowable error. From this you know each point is OK to use for control.

Elevations and topographic features in this part of the field notes shown by offsets from ¢ of Birch Tree Dr.

		X - SECTION			
		BIRCH	TREE	DRIVE	
	0+00 = 100'	S/O: SOUTH ℄			
STA	LT.	℄	15'	RIGHT 20'	
0+75		109.12 ℄	108.82 15' E.P.	111.32 20' ℄	112.43 30'
+93			WATER VALUE 13.5		F.H 22.1'
1+00		110.12 ℄	109.82 15'	112.11 20' ℄	113.52 46.5
+15					6" ⌀ ACACIA. 62.1
+25		111.12 ℄	110.82 E.P.	112.80 ℄	113.97 45'
+50		12.12 ℄	11.82 E.P.	13.50 ℄	14.73 45'
+57					3" ACACIA .62'
+75		13.11 ℄	12.83 E.P.	14.19 ℄	15.23 45'
+94					4" BIRCH 38.3
2+00		14.12 ℄	13.81 E.P.	14.82 ℄	15.79 45'
+14			WATER METER 21'		15.68 TOP
+25		15.13 ℄	14.82 E.P.	15.93 ℄	16.97 45'
+40					5" BIRCH 40.5'
+42					3" BIRCH 31.2
+47					6" BIRCH 37.9

STA.	LT.	₵	RIGHT 15'	20'	3
2+50		16.12 ₵	15.82 E.P.	16.82 R	18.11 / 45'
+75		17.10 ₵	16.80 E.P.	17.75 R	19.22 / 45'
+87		12.60 ₵	17.68 16.5' E.P.	18.27 R	
+93	S.M.H. 5'	17.84 E INT.	17.78 17'	SEE DETAIL BELOW	
3+00		18.12 ₵			
+07			W.V. 13.8'		
+25		119.12 ₵	118.82 15' E.P.		

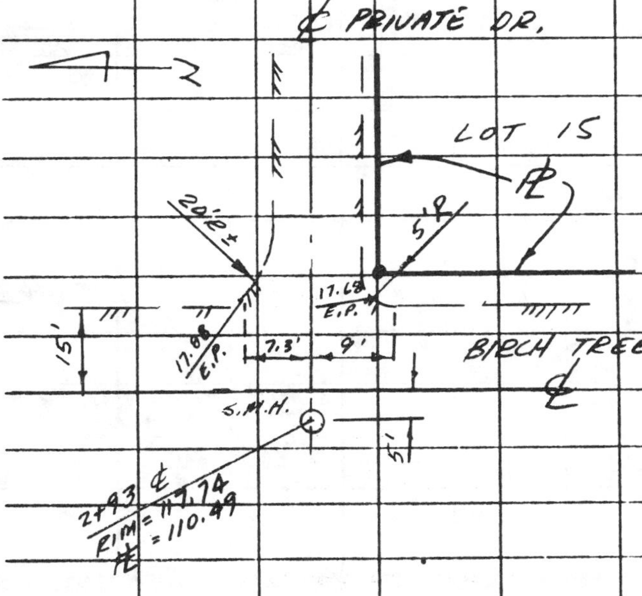

₵ PRIVATE DR.

N

LOT 15

PL

20'±

17.68 E.P.

17.68 E.P.

7.3' 9'

BIRCH TREE

₵

15'

5'

5'

S.M.H.

2+93 ₵ = 119.74
RIM = 110.49
PL

Note sketch to clarify offsets and controls for pavement outline

Elevations and topographic features in this part of the field notes shown by offsets from of the private driveway beginning at ¢ Birch Tree Dr.

STA.	LEFT 10'	5'	¢	RIGHT 5'	6' ℞
	X- SECTION				
	PRIVATE DRIVEWAY				
	1+00 = ¢ INT BIRCH TREE w/¢ DRIVE				
0+95			S.M.H. ¢		
1+00			117.84 ¢		
+17			117.78 ¢		
+25	119.02 10'GRD	118.19 5'E.P.	118.10 ¢	118.20 5'E.P.	18.60 6'℞
+26				P.P. #1YE3461 5.5'	
+50	20.13 10'GRD.	19.45 5'E.P.	19.35 ¢	19.44 E.P.	19.93 ℞
+73					
+75	21.35 10'	20.71 E.P.	120.60 ¢	20.19 E.P.	20.55 ℞
+77					
+80					
+85					
+87					
+91				P.P. #1YE3462 5.5'	
+92					
+93			S.M.H ¢	RIM= 121.50 ℞ = 114.37	
2+00	22.12 10'	21.83 5'E.P.	21.85 ¢	21.96 5'E.P.	21.62 10'GRD.

Note elevation of both rim o[f] sewer manhole at surface of street and at the flow line.

4

NOTE: NO EVIDENCE ON SURFACE
      OF EXIST. GAS LINE. IMP.
      PLAN FOR BIRCH TREE
      INDICATES 3" GAS 17'
      E/O ₵ BIRCH TREE w/
      SERVICE CON. @ 10'± N/O
      DRIVEWAY.

	19.28				
	20' GRD.				
4" BIRCH					
11'					
	19.82				
	20'				
	2" BIRCH				
	20'				
3" BIRCH					
10'					
	3" BIRCH				
	20'				
2" PINE					
16'					
		6" PINE			
		29'			
6" PINE					
13'					
	120.38				
	20'				

Notes on this
side cover
information
inside of Lot
south of ₵
private drive.

Bench mark
elevation from
level circuit. ⟶

This note tells
you that 0°
Azimuth was
set on Station
"B" and angles
were turned to
the left.

This shows
transit set at
Station "C".

These notes
were taken by
stadia rod
except as noted.
Note ground
elevations
taken and
shown to tenths
of a foot by
stadia which is
accuracy
comparable to
the conditions.

STA. AZIM. 0°00'00" B.S. "B" ∡'s LEFT	+ B.S.	H.I.	F̄.S.	ELEV.
⊼ ℮ "C"				117.93'
0°00'	5.87	123.80	2.75	121.05
0°00'			3.5	120.3
354°35			3.4	120.4
0°00'			4.3	119.5
4°40'				
6°10'			4.3	119.5
9°05'				
14°30'			4.3	11.9.5
22°10'			5.0	118.8
27°45			5.5	118.3
32°20'			6.1	117.7
43°30'			7.3	116.5
36°45			6.5	117.3
43°30'			6.5	117.3
27°45'			5.9	117.9
20°25'			5.2	118.6

(CONT ON PG 6)

SI / HOR. DIST.	VERT.∡	VERT. DIF.	ELEV.	NOTES 5
				GRD. @
				2x2 @ "C"
1.265 / 126.5				2x2 @ "B"
1.05 / 105.0				GROUND @ ℞
0.95 / 95.0				GRD.
0.73 / 73.0				GRD. @ ℞
0.875 / 87.5				4"∅ FIR
0.94 / 94.0				GRD.
0.98 / 98.0				6"∅ FIR
1.195 / 119.5				GRD.
1.25 / 125.0				"
1.175 / 117.5				"
1.11 / 111.0				"
1.04 / 104.0				"
0.94 / 94.0				"
0.61 / 61.0				"
0.79 / 79.0				"
1.00 / 100.0				"

Stadia Intercept (SI) times 100 equals distance.

This form has 2 sets of columns for elevations. The left hand page is for readings with the telescope level. The right hand page shows vertical angle and difference of elevation.

STA	HOR. ∢	+ B.S.	H.I.	F̄.S.	ELEV.
	15°50′			5.2	118.6
	14°45				
	26°32′			6.0	117.8
	19°25′				
	11°25′			5.3	118.5
	0°00′			4.7	119.1
	352°40′			4.4	119.4
	346°05′			5.3	118.5
	0°00′			5.1	118.7
	26°32′			6.0	117.8
	47°32′				
	33°30′			6.4	117.4
	61°02′			6.9	116.9
	66°31′				
	52°40′				
	57°05′				
	73°32′			7.4	116.4
	60°50′			7.2	116.6

(CONT. ON PG. 7)

S1 HOR. DIST.	VERT. ∢	VERT. DIF.	ELEV.	NOTES 6
0.82/82.0				GROUND
0.71/71.0				5" ∅ FIR
0.58/58.0				GRD.
0.51/51.0				3" ∅ FIR
.54/54.0				GRD.
.535/53.5				GRD. @ ℄
0.57/57.0				GRD.
.18/18.0				"
.35/35.0				GRD. @ ℄
.18/18.0				GRD.
11.0				3" FIR MEAS. W/TAPE
0.30/30.0				GRD.
0.236/23.5				"
0.31/31.0				2" ∅ BIRCH
0.36/36.0				6" ∅ BIRCH
.44/44.1				4" BIRCH
0.39/39.0				GROUND
.515/51.5				"

The field man takes elevations at the critical points or "breaks" of the ground; therefore, the slope between these points is considered to be constant.

STA.	HOR. ∡	+ B.S.	H.I.	‾ F.S.	ELEV.
	54°01′			6.9	116.9
	49°55′			7.4	116.4
	60°51′			8.2	115.6
	70°15′			7.5	116.3
	75°20′			7.5	116.3
	92°55′			7.9	115.9
	82°50′				
	90°02′		123.80		
	96°30′		123.80		
	112°35′		″		
	117°30′		″		
	106°40′				
	99°35′		123.80		
	108°01′				
	117°30′		123.80		
	227°05′			8.1	115.7
	121°31′			7.9	115.9
	127°02′			7.6	116.2

(CONT. ON PG 8)

7

SI HOR.DIST.	VERT. ∢	VERT. DIF.	ELEV.	NOTES
.57 / 57.0				GROUND
0.835 / 83.5				"
0.88 / 88.0				"
0.680 / 68.0				"
0.58 / 58.0				"
.515 / 51.5				"
.645 / 64.5				4"∅ ACACIA
0.81 / 81.0	−6°05'	−8.58	115.2	GROUND
.965 / 96.5	−5°39'	−9.50	114.3	"
0.91 / 91.0	−5°48'	−9.20	114.6	"
0.920 / 92.0	−5°44'	−9.20	114.6	"
0.73 / 73.0				4" ORANGE
.620 / 62.0	−7°41'	−8.30	115.5	GRD.
0.53 / 53.0				4" ORANGE
0.65 / 65.0	−7°25'	−8.40	115.4	GRD.
.615 / 61.5				GRD @ ℄
0.45 / 45.0				GRD.
0.37 / 37.0				"

Because of difference of elevations and trees in the way, the vertical angles with rod readings were used here instead of direct rod readings. Reductions were made by the sin-cos formula.

STA.	HOR. ∢	+ B.S.	H.I.	− F.S.	ELEV.
	103°33′				
	92°28′			7.2	116.6
	227°05′			7.1	116.7
	121°31′			6.6	117.2
	49°1T′				
	132°55′	∢ FROM PT. "B" TO PT "D"			
		∢ CHECKS OK			

This angle to the SW corner of the water meter is a "positional" check for benefit of the draftsman.

Note final check from "B" to "D" to verify stability of setup at "C".

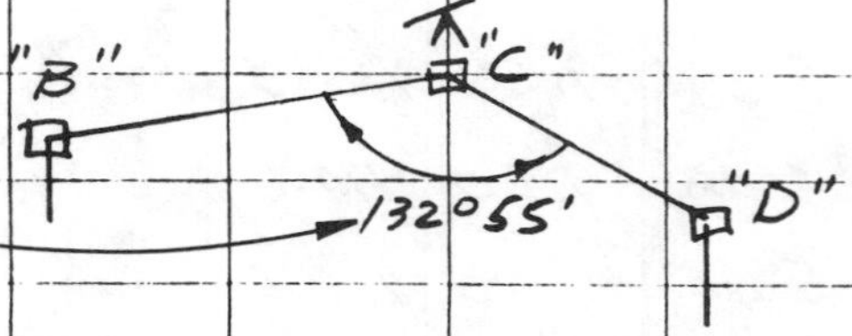

"B"

"C"

"D"

132°55′

8

S1 HOR. DIST.	VERT ⊄	VERT DIF.	ELEV.	NOTES
.315 / 31.5				6" ⌀ BIRCH
0.25 / 25.0				GRD.
0.26 / 26.0				GRD. @ ℞
.135 / 13.5				GRD.
				S W cor. W.M.

In conclusion, let us realize that although a representative group of examples of field notes have been presented here, there are many more different forms used in the more specialized areas of surveying. If your work carries you into those avenues, you can, and you should, get samples of their forms and study them.

Abbreviations

In the rush of note keeping and the limited space available for the plethora of information, abbreviations are a must; consequently, it is necessary to have some standardization so that all persons involved will use and interpret them with the same meaning. Whereas the list given in Appendix A has been developed especially for application to civil engineering and architectural projects to which the surveyor addresses himself, there are more extensive ones available stressing subject matter in particular areas such as Technical Manual 5-581A from Headquarters, Department of the Army.

Topographic Symbols

There are many symbols used by government agencies and private practitioners, most of which bear a close resemblance to the physical features which they are intended to represent. Those shown in Appendix B are a good representation of the most commonly used.

PROBLEMS

1. What do "field notes" represent in general?
2. Compared with a "boundary line" survey, what does "topography" generally cover?
3. What differentiates "engineering" surveying from "boundary" surveying?
4. Where and why would you use "angle intersections" instead of direct ties in field notes?
5. When would field notes be drawn in the office?
6. For what type of project is a "control" survey used?
7. What is the value of "tie lines" recorded in the field notes?
8. What is the difference between a "precise" control survey and a "stadia" control survey?
9. What are 4 operations included within the scope of "geodetic" surveying?
10. What are the 2 principle methods of determining the position of a boat for hydrographic soundings?
11. Why is an accurate traverse needed on the inside area of a project?

Chapter *2*

Plotting Traverses

Methods*

Whether you are given field notes from a topographic survey with any number of stations to plot for development of the information obtained or a survey of a long strip of land for a new right-of-way acquisition to be related to the surrounding properties, each one

---

*Because there are any number of books on surveying that teach the mathematics of traversing, it is assumed that you have had those rudiments and therefore the explanation will not be duplicated here.

requires that the lines traversed by the survey be plotted on paper. The method you use will be determined in part by the manner in which the man in the field did his work.

Use of Traverses

There are in general two types of traverses, open and closed. The open type starts at a significant point which may be either an established point with controls or just one related only to the project, and then proceeds to delineate the strip or area under surveillance.

The closed traverse may also start at either a controlled point or an arbitrary point but it then proceeds by way of distances and angles until it returns to the point of beginning.

There are several different ways of plotting a traverse. The angular relation between successive lines may be measured and plotted by deflection to the right or left from the forward prolongation of the previous line; the angles may be measured and plotted on the inside (called interior angles) of the traverse; they may be measured and plotted on the outside (called exterior angles) of the traverse, or by azimuthal angles from a stated line. In each case, of course, the distance is measured along the related line.

Deflection Angles and Distances

If the field notes show the successive courses with a deflection angle turned from the forward projection of the previous line, and of course the distance with it, as shown in Figure 2-1, then the simplest way to plot it is to follow the same procedure. First, however, check the angles by adding the *algebraic* sum of the deflection angles (right = +, left = −) which should = 360°.

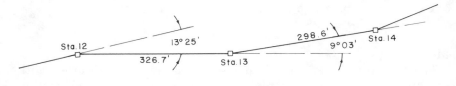

Fig. 2-1

Using the protractor with the center set on the station and 180 degrees along the *back* line gives you 0 degrees ahead from which the deflection angle can be laid off. Draw a line on the deflected angle and scale the distance along it to establish the next station.

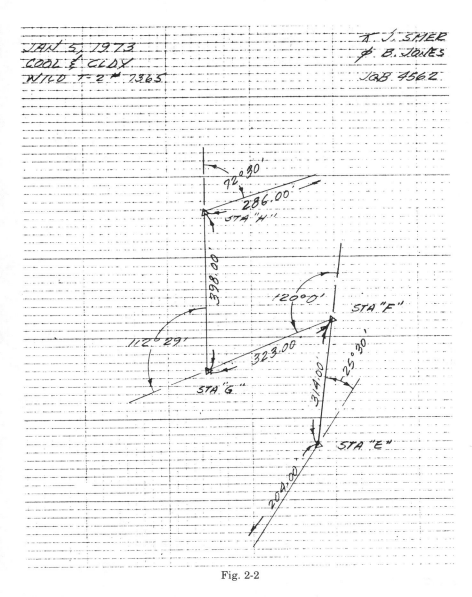

Fig. 2-2

EXERCISE 2-1: Plot the five courses shown in the field notes in Figure 2-2 with protractor and a scale of 1″ = 100′.

* * *

If it is suitable, sometimes the field notes are drawn to scale but usually not, so do not rely on the field man's drawing for scaling either angles or distances.

Another way to plot deflection angle lines is by the application of natural tangents. Although this is more accurate, it is not as commonly used because it requires a calculation for each line and is slower than using the protractor.

There are two ways to use the tangent method. One is to set off 10 units of your map scale (or 10 inches if the other is too short for good accuracy) on the prolongation of the previous line measured from the last station point; multiply the tangent of the deflection angle by 10 and construct that right angle distance from the 10-unit point and connect it with the last station as shown in Figure 2-3.

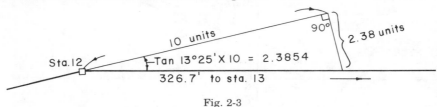

Fig. 2-3

The other way is to multiply the distance between two stations by the cosine of the deflection angle to determine the position of the right angle point on the projected line which is opposite the next station, then multiply that distance by the tangent of the deflection angle (or multiply the measured distance by the sin of the deflection angle) to get the offset to the station point. See Figure 2-4.

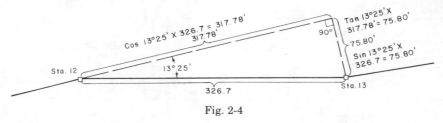

Fig. 2-4

With electronic computers, very little time is required for either one of these tangent methods.

EXERCISE 2-2: Apply each tangent method explained above to plotting the line between Stations E and F in Figure 2-2 and compare it with the line drawn with the protractor.

Bearings

In the above example, the lines created by the deflection angles could also be translated into lines with bearings and the traverse plotted by using a drafting machine. A bearing is the direction of a line normally expressed in symbols of one of the four quadrants of the compass in conjunction with the angle expressed in degrees, minutes and seconds; i.e., N 32° 46′ 21″ E. The angle is usually based on a previously established base line although in some cases, it is developed from a true (which means astronomical) north line established by either a polaris or solar observation.

To establish the bearing on paper, the angle is scaled either from 0° North or 0° South and is limited to not more than its 90° quadrant.

Although S 32° 46′ 21″ W has the same angular position on the paper as N 32° 46′ 21″ E, the direction of travel along the line is opposite. See Figure 2-5.

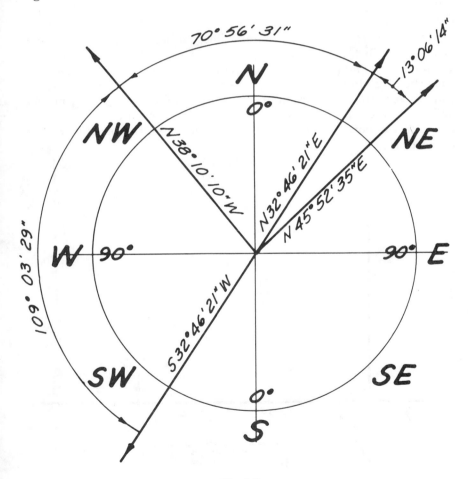

Fig. 2-5

When angles in the two north quadrants or the two south quadrants straddle the 0° line, the total angle is obtained by adding the *algebraic* sum of the two bearings. Again referring to Figure 2-5, the angle between N 38° 10′ 10″ W and N 32° 46′ 21″ E is the sum of the two which equals 70° 56′ 31″. The angle between N 32° 46′ 21″ E and N 45° 52′ 35″ E is the algebraic sum or 45° 52′ 35″ + (− = descending) 32° 46′ 21″ which equals 13° 06′ 14″. When angles cross the line between a north and south quadrant, the two bearings in the same north or south quadrant are added together and their sum is subtracted from 180° 00′ 00″. In the example, N 38° 10′ 10″ W plus N 32° 46′ 21″ E equals 70° 56′ 31″ and subtracting this from 180° 00′ 00″ gives 109° 03′ 29″ which is the total angle in the two west quadrants.

To convert deflection-angle lines into bearing lines, you must establish the first line used in the survey with a bearing developed from another line from the same point of beginning which has a known bearing. For instance, if the survey starts at the southwest corner of Lot 4 (Figure 2-6) and proceeds with a deflection angle of 21° 32′ northerly from the north line of 1st Street, which has a bearing of N 89° 41′ 10″ E, you convert the angular difference to a bearing by subtracting the deflection angle from the street bearing. In other words, the new line into Lot 4 is northerly from the street; therefore, the bearing will be closer to North or

$$
\begin{array}{r}
\text{N } 89°\ 41′\ 10″\ \text{E} \\
-\quad 21°\ 32′\ 00″ \\
\hline
= \text{N } 68°\ 09′\ 10″\ \text{E}
\end{array}
$$

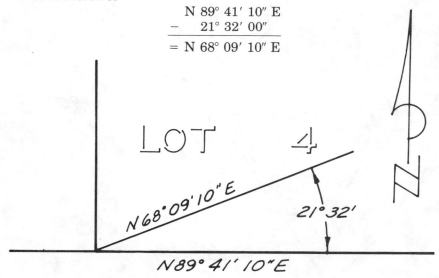

Fig. 2-6

Having thus established a base, the bearing of each succeeding line will be determined by adding or subtracting the deflection angle from the previous bearing. With a bearing assigned to each line, you can then set your drafting machine on the base line and proceed from there with bearings and distances.

Full Angles

Sometimes the field notes report the angles as interior or exterior instead of by deflection. With a closed traverse, you should total the interior angles to determine if you have a mathematical closure before you draft it. The formula for determining the total number of degrees contained within a polygon is $(n-2) 180°$ in which n is the total number of angles in the traverse; therefore, if you have 7 angles, the sum of all the angles should equal $(7-2) 180°$ or $900°$.

If some of the recorded angles are exterior, you will need to subtract them from $360°$ to get the matching interior angle. These are usually plotted with a whole-circle protractor unless you translate the positions of the lines into bearings for plotting with a drafting machine.

Azimuth Angles

In rare cases, field notes are made from angles which are all turned from one azimuthal point which may be north or south or an assumed point or line. In plotting from these notes you keep the $0°$ of the protractor always in the same direction as the azimuth which was used in the field. This operation can be performed for each successive course or a group of angles can be marked around the protractor at one time and then the lines transferred by parallel lining to their respective course positions with the appropriate distance.

If your drafting machine has the full circle $360°$ indexing, it can be used in the same manner; however, if it is graduated only in each of the four $90°$ quadrants, you will need to convert the azimuth angles to bearings first in order to use it.

Rectangular Coordinates

The desirability of this method of plotting is that each course is independently established and not dependent upon the combined drafting of all the previous courses which could result in cumulative errors.

The principle of this method is that a grid of control lines is drawn on the paper at intervals of a large number of whatever scale unit is being used; i.e., if the scale is $1'' = 100'$, the grid may be set at $200'$, $500'$ or $1,000'$ intervals, or, if the scale is $1 \text{ cm} = 10 \text{ m}$, the grid may be set at 50 m or a hectometer. The zero control is usually set to the bottom and left (south and west) of the drawing area so that all coordinates are plus north and plus east. This is even true with using State Plane Coordinates.

Before plotting coordinates, the traverse should be checked for mathematical closure and adjusted if necessary. If one of the points in the field has been tied into another point with known coordinates, this will control the traverse and all other points will be assigned coordinates derived from calculations. If there is no tie to any established coordinates, then one point can be given a pair of even numbers (or any assumed numbers) from which to start.

The grid lines should be laid out on the paper as quickly as possible so as to minimize any difference in the subsequent scaling due to shrinking or stretching of the paper caused by atmospheric conditions. If the paper succumbs to any appreciable amount of change, you will need to determine the amount and apply a proportionate correction to your scaling.

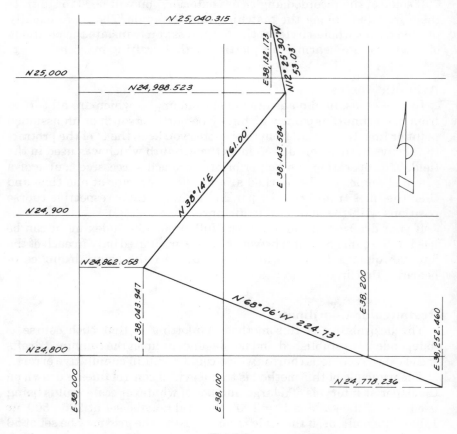

Fig. 2-7

Having derived coordinates for each point of your traverse, you establish their positions respectively along the North grid lines and East grid lines, then project *each pair* to its intersection as shown in Figure 2-7. After connecting successive points, you can check the bearing and distance between them for verification against the traverse.

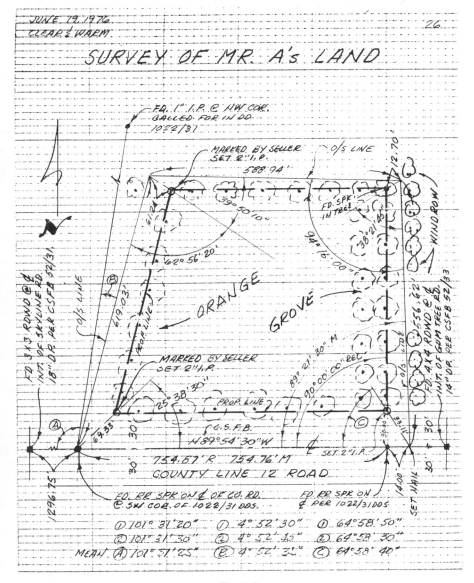

Fig. 2-8

Offset Points

In some cases physical conditions prevent the survey crew from occupying the corners or lines which are to be measured; consequently, offset lines must be run with tie lines to the true corners from which calculations are made for the delineations of the property lines to be shown on the map.

EXERCISE 2-3: Mr. A has an agreement to sell that portion of his orange grove outlined with a heavy dash line on the field notes shown in Figure 2-8.

From the information shown in the survey notes, calculate the bearings and distances of the property lines and draw a map relating them to County Line 12 Road and the westerly line of the land described in 1022/31 of Deeds. You can do this by either forced closures or coordinates. Do NOT show offset lines on your map.

Here is a situation where coordinates can also be applied to good advantage.

*   *   *

Scales

The first part of this chapter has been primarily a discussion of angles, but since distances are a conjunctive part of mapping, we must consider the means of transferring the field measurements onto paper. This is accomplished by the use of scales, each of which is in effect a constant proportionate reduction from the actual ground measurement to an appropriate measure for the drawing.

Although the triangular scale is common for architectural and engineering uses, flat scales are also used and particularly with drafting machines. The triangular form is available in two shapes, flat ▲ or concave ⚎, while the flat forms have either two bevels ◢━━━, opposite bevels ◢━━━◣ or four bevels ◆━━━◆. Drafting machines use the two bevel form scales with chuck plates for quick attachment. See also Chapter 5 on Drafting Tools.

Architectural

This scale is available in both triangular and flat-with-bevel forms. In the triangular form scale, one edge is cut into 12 inches, the same as a regular ruler with the smallest division being 1/16″. The remaining five edges each have two scales paired as follows: 1/8″ and 1/4″, 3/8″ and 3/4″, 3/32″ and 3/16″, 1/2″ and 1″, and 1½″ and 3″. Each unit is divided into parts of a foot at its end of the scale and the remaining marks are whole units.

The 1/8″ scale reads from LEFT to RIGHT while the 1/4″ scale reads from RIGHT to LEFT. The 1/8 is interpreted as 1/8″ = 1′ or 1″ = 8′; except for the subdivision of the 1/8″ into 6 parts to the *left* of 0, all the other marks indicate one foot and note that the foot numbers are at every fourth mark and are shorter than the ones for the 1/4″ scale.

The 1/4 is interpreted as 1/4″ = 1′ or 1″ = 4′; except for the subdivision of the 1/4″ into 12 parts (each for one inch) to the *right* of 0, the other *long* marks indicate one foot and note that the foot numbers are at the alternate long marks and are a larger size as shown in Figure 2-8.

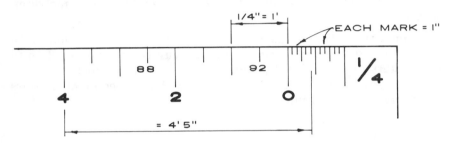

Fig. 2-9

Thus, if you are to draw a house on paper so that 1″ on the paper = 4′ on the ground, you use the 1/4 scale.

Whatever unit of scale you choose to use, you will find the subdivided part at one end of the ruler with the whole unit marks reading toward the other end. The same principle of the foot unit representation plus the foot-divided-into-inches unit beyond the zero point at the end applies to whatever scale you use.

The most commonly used scales are 1/4, 1/8 and 1/16 for building plans while the larger ones are used for detail work.

Engineering

This scale is also available in both the triangular and flat-with-bevel forms. The triangle form carries six scales which are usually divided into 10 parts to the inch, 20 parts to the inch, 30 parts to the inch, 40 parts to the inch, 50 parts to the inch and 60 parts to the inch. It is also possible to get a triangular scale cut into 80 parts to the inch in lieu of the 10 parts. The flat form scales are available in the same gradations.

The 10 parts to the inch scale can be used as 1″ = 1′ or 1″ = 10′ or 1″ = 100′ or 1″ = 1000′. The other scales can likewise be used in any multiple of their indicated number. For special work, the 80 and 100 scales are available with full 80 parts and 100 parts to the inch marked on the face.

For special survey work in section land where the chain measurement of 66 feet is used, there are special scales divided into $1'' = 66'$.

Another scale used by engineers is divided 4 inches = 1 mile in 20 foot increments and 2 inches = 1 mile in 100 foot increments.

Metric

These scales are available in the triangular form or in the flat form with either two bevels or opposite bevels. The flat forms are available in a combination of inches on one side and metric on the other or metric on both sides.

In similarity to the engineer's scale, they have marked divisions in 10 cm., 20 cm., 30 cm. and 50 cm. groupings with centimeters reading to millimeters or to half-millimeters. Some are marked in the proportionate manner; i.e.; 1:1, 1:2, 1:5, 1:10 or 2:1 (2x) divided in millimeters and numbered in millimeters.

Topographic

These drawings are produced using a variety of scales depending upon their form of creation and final use. In the development of architectural plans, the architect may want a ⅛″ or ¼″ to the foot scale topo map. The engineer may prefer a 100′, 400′, 500′ or a 1,000′ to the inch scale map or a 30 m, 50 m or 100 m to the centimeter scale map.

Topographic maps prepared from air photos are made on scales expressed as a fraction or a ratio; these are also known as natural scales. For instance, a ratio of 1:62,500, or fraction of 1/62,500, means that 1 inch on the map denotes 62,500 inches on the ground, 1 foot on the map denotes 62,500 feet on the ground, etc., etc. This ratio, or fraction, is usually found at the bottom of quadrangle maps and sometimes designated along with the graphic scale.

Very often a graphic scale is shown along with the expressed scale so that if the map material shrinks or expands in the reproduction processes, or due to atmospheric conditions, the scale, which will also change in a similar amount, will still be applicable and more adaptable than the expressed scale.

Graphic scales as shown in Figure 2-9 are drawn to the actual lengths with whole units to the right and one unit divided into parts on the left. These are more often used on topographic maps produced through photogrammetric means rather than on architectural drawings or survey produced maps; however, anytime that a map may become subject to reduction, or expansion, by reproduction for uses in reports or other forms of display, a graphic scale is most desirable.

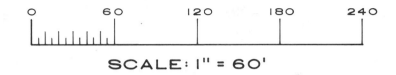

SCALE: 1" = 60'

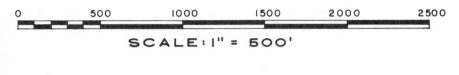

SCALE: 1" = 500'

1/62,500

1/62,500

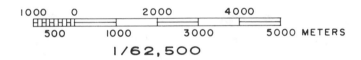

1/62,500

Fig. 2-10

PROBLEMS

1. Name 3 methods of laying out deflection angles.
2. How would you determine the angle between one bearing in the northeast quadrant and the other bearing in the southeast quadrant?
3. Name 3 methods of plotting the angular relationship between successive lines in a traverse.
4. What are the 2 procedures for checking a set of angles in the field notes before you plot them?
5. What is the advantage of plotting by rectangular coordinates?
6. How do you check the coordinates *before* using them for plotting?
7. What are the 3 types of scales which you will probably use in drafting?
8. Explain a topographic scale.
9. Explain graphic scales.

—NOTES—

Chapter 3

Relating Maps To Descriptions

Legal Descriptions

    Because many maps have to do with property boundaries, it is necessary to have some comprehension of legal descriptions in order to follow and draw the outline of the area within which the pertinent information must be placed.

    Reference to a piece of land already outlined on a map usually requires only copy work but when it, or any other parcel, is cut into parts by a metes and bounds description, you need to understand the legal value of different words and phrases in order to know what to do with any contradictory statements in the description or differences between the figures in the description and the surveyor's field measurements.

Monuments found by the surveyor may or may not be acceptable for control. Your proper interpretation of the description will determine the acceptability of the monuments.

Lots or Parcels On A Map

When you are given field work in note form to draw into and around a parcel of land, the outline of which you find is already shown on an officially recorded map, the presumption is that you merely copy the measurements, angles and ties shown on that map. Normally this is satisfactory, BUT, if the surveyor finds discrepancies between monuments in the field and measurements on the map, be sure *you* show those differences; do not assume they were intended to be the same and neglect showing your client the facts. It may even be appropriate to set off an enlarged detail of the problem area.

When such differences occur around a single lot or parcel, considerably more information will need to be shown in the surrounding area to explain the reason for the alteration. This will be treated more specifically in the discussion on American Land Title Association (ALTA) maps in Chapter 11.

Government Section Land

This kind of land is also known as Public Land or Township Subdivisions. The full and correct designation is the United States System of Rectangular Surveys. What you must be most careful about is that although the *theoretical* section is one mile square and the *theoretical* township is six sections east and west by six sections north and south with perhaps a little less than one mile east and west in the tier of sections along the west side of the townships (which variation was planned to accommodate the convergence of the meridian) and likewise a little less (or perhaps a little more) than one mile north and south in the row of sections along the north side of the township; in actuality, it is not always found to be that way. One area in Ohio has townships with five sections east and west by five sections north and south. Also, in those areas where one meridian bounds another, the townships are very often fractional and for purposes of differentiation from standard are usually designated as fractional, for example, T 34½ N.

In order to understand why these commonly expressed concepts of even-mile section lands are not acceptable, one must look into the field work from which the maps were made.

The great majority of the first township plats submitted, and approved, in every area of government surveyed lands conformed to the prescribed format. It was when new township plats were made from *re*surveys that discrepancies were clearly illustrated showing the results of the earlier field work. Some of the odd shapes and sizes of sections are difficult to believe, but for the facts shown on the plats and in the field notes of which an example is shown in Figure 3-1. If you have not already seen some of these, no doubt you can find examples in the files of your particular District Land Office.

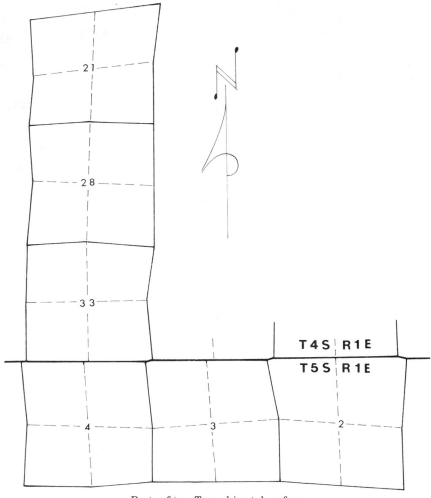

Parts of two Townships taken from
Official Approved Plats of Resurveys.
Fig. 3-1

It is an extremely rare case when a section line between found original monuments or between correctly reestablished corners, measures exactly 5280 feet or 2640 feet to the quarter section monument. More commonly, there is found a difference in measurement all the way from the best part of a foot up to many feet and sometimes more. In fact, one particular Section 6 is a mile and one-half east-west, while in another area, a certain Section 3 is over a mile and one-half north-south. This is to point up the fact that you should be fully aware of these possibilities and not expect a description of "the west half of the southwest quarter of Section 12" to necessarily yield a 1320 foot by 2640 foot parcel of land. For this reason, the layman should be apprised of the fact that he should not rely on subdividing the section on paper because in all probability a resurvey on the ground in conformance with the rules will disclose any number of differences.

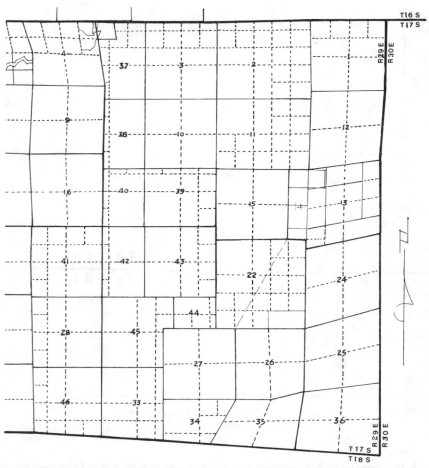

Easterly portion of T 17 S, R 29 E According to Official Approved Plats
of Surveys and Resurveys.
Fig. 3-2

A different reason for some of the large discrepancies is that separated areas have been surveyed without having first established the entire framework of township lines; consequently, there were voids containing excessive distances which were unknown until subsequent surveys filled them in. This is why one township in particular shows 46 sections even though one is missing and another one is questionable. See Figure 3-2.

The rules set forth in the Manuals of INSTRUCTIONS FOR THE SURVEY OF PUBLIC LANDS in the UNITED STATES tell how to subdivide sections, whole or fractional, and even though they were contemplated on the basis of a regularly developed section, the rules still apply even if the section be distorted.

For example, Figure 3-3 shows a config-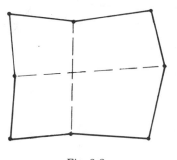
uration resulting from found monu-
ments but still the division into halves
and quarters must follow the rule of
joining opposite quarter corners with
straight lines. Many examples of odd-
ities in sections can be cited but the
point to remember is, follow the rules in
every way possible based upon the most
current survey showing measurements
and angles between *acceptable* found   Fig. 3-3
monuments.

The numerical order for townships,
ranges, sections and lots is explained in
Sections 3-1 through 3-84 in the 1973 Manual. When a section does not contain the regular 640 acres, as in the west tier or north row in a township, the odd areas are assigned lot numbers. Also, when sections are cut up by a river, a lake, a rancho, etc., the land between the regular areas (usually down to 40 acres) and the irregular line is cut into lots. Figure 3-4 shows four typical examples from the Manual.

Lots in sections on an approved township plat are "whole" lots, not "fractional" lots. The very reason they exist is because the area so designated is a fractional part instead of a regular part (40 ac, 80 ac, etc.) of a section.

The word "fractional" is properly applied to those *sections* cut off in one or more parts by reason of an irregular boundary of water or straight lines of private grants.

Unless your surveyor has actually done enough field work by which to calculate the boundaries of a lot for you to draw, your next best proce- dure is to simulate the showing on the township plat. A simple example of a section land description from which you may have to draw a map is thus:

> "Lot 1, Section 4, Township 3 North, Range 2 East, of the First Principal Meridian, in the County of Oshkosh, State of South Dakota, according to an official plat of said land filed in the District Land Office."

You should begin by studying the township plat and then take the most recent survey information available concerning monuments and measurements from which to draw your map.

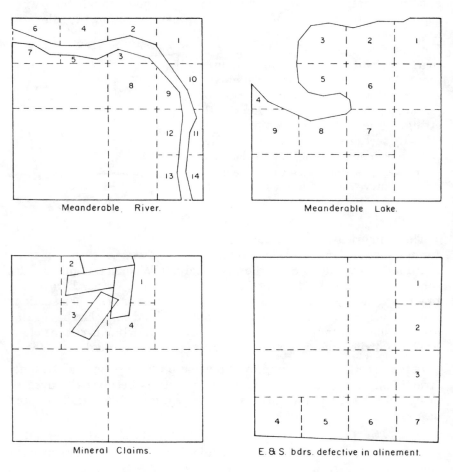

Examples of subdivision of fractional sections with lots.
MANUAL OF SURVEYING INSTRUCTIONS
BLM Technical Bulletin 6
Fig. 3-4

If you have a portion or portions of a section which are designated specifically, such as "The southwest quarter of the northeast quarter of Section . . . . ," you have a definite part of the section described and do not need "described as follows: ." If there is only a part described by metes and bounds, the form will be thus:

"That portion of the southwest quarter of the northeast quarter of Section . . . . ."

continuing with:

"described as follows:"

In this case, you lay out the survey of the metes and bounds part and show all the ties to the section, or sub-section, lines, corners and monuments. An example of this is shown in Figures 1-4a and 1-4b field notes.

Another matter concerning lots in a section subdivision is that of water boundaries. The meandering line of the lot along the ocean, lake, river, creek or stream is a survey line for the determination of area, while the line of *title ownership* follows the high water mark, the low water mark, or the thread of the stream, according to whatever determining conditions exist. You normally will draw your map showing the meander line as surveyed. If you have enough information, you can *indicate* the water line.

Sometimes a parenthetical phrase is inserted or added and this information can be helpful because it clarifies the facts.

A simple type of parenthetical description would read thus:

The west half of the northeast quarter of the northeast quarter of the northeast quarter of Section 10, T . . . . etc., described as follows:

Beginning at a point on the north line of said Section South 88° 42′ 55″ West 328.78 feet from the northeast corner of said Section; thence continuing along said north line South 88° 42′ 55″ West 328.78 feet; thence South 0° 23′ 24″ West 662.18 feet; thence North 88° 36′ 48″ East 329.02 feet; thence North 0° 01′ 36″ East 661.96 feet to the point of beginning.

This indicates that a refined survey of the section had been made and that the bearings and distances cited reflect the true ground conditions. These would hold over the fractional reference because they are more definite. The shapes of thousands of government-surveyed sections as shown on township plats have been changed as a result of subsequent surveys finding the original monuments awry.

EXERCISE 3-1: Draw the above description to a scale of 1″ = 40′ or convert the feet to metric and draw on a 1:20 scale. Note the difference between this and a theoretical 330′ x 660′ rectangle.

* * *

When a description is written with good intent but without sufficient qualification, it creates a question as to the *specific* intent. Consider the lines shown in Figure 3-5 where the description in the deed read, and without a caption,

> "Beginning at the southwest corner of Section 11, T__ __, R __ __, of the _____ Meridian; thence North 10 chains, East 10 chains, South 10 chains, West 10 chains to beginning."

Fig. 3-5

Again this apparently was on the presumption of a *regular* north-south and east-west section. When such language is used in this manner, it has been held that such terms, lacking any qualification, are construed as equivalent to, and based upon, an astronomically established north on a meridian of longitude and the "east" and "west" lines are considered as straight lines at right angles to that meridional line.

From the drawing, the obvious problem is lack of conformity between the described lines and the section lines. If there had been a caption of "That portion of Section 11 . . . . etc.," the triangle included in Section 10 would be automatically cut off and there would be the triangular gap along the south line of Section 11.

The point to remember here is that if you are confronted with such a problem, don't offer to solve it, explain it and request enough information to make it workable or leave it until you get more survey dope.

To show you what would be desirable, if the description had been written not only with a caption but with each line properly qualified, the result would have given a parallelogram conforming with the boundaries of the section. A correct form would read thus:

> "That portion of Section 11, T __ __, R __ __, of the _____ Meridian in the County of _____, State of _____, according to the official plat of said land filed in the District Land Office described as follows:

> Beginning at the southwest corner of said Section; thence north 10 chains along the west line thereof; thence east 10 chains parallel with the south line thereof; thence south 10 chains parallel with the west line thereof to the south line of said Section; thence west 10 chains to the point of beginning."

This wording would hold the west and south lines coincident with the section lines in spite of the fact that they differed from "true" lines as much as 8° and 6° respectively.

A short form description which would accomplish the same objective would read thus:

> "The southerly 10 chains of the westerly 10 chains, measured along the west and south lines respectively, of Section 11, T ___ ___, R ___ ___, of the _____ Meridian in the County of _____, State of _____, according to the official plat of said land filed in the District Land Office."

Here is another problem of which you need to be cognizant so that you do not draw your map on the assumption of coincident lines but *show the true facts.*

Some descriptions of section land not only tie to and along the section (or sub-section) line but also to adjoining owner's land. This can introduce the problem of complicity from excess ties if the adjoiner's line is not coincident with the section (or sub-section) line. For instance, if the description of the land on the east side of the sub-section line shown in Figure 3-6 ties also to the "east line of Tom Brown's land on the west," it can raise the question of which one in fact will control. Extensive research back through the chain of title may then be necessary to determine the validity and choice of references.

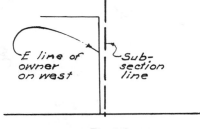

Fig. 3-6

When title is held within a quarter section or quarter-quarter section or some other definitive part of a section, there is no need of making a tie to the adjoiner's land by the same analysis that when you refer to all or part of a lot in a subdivision, you do not tie to John Jones' land in the adjoining lot because the lots have been established by survey and have common lines so your reference to the lot according to the map is sufficient. Likewise, when the sub-section lines are established by survey and/or the adjoiner's deed ties to the same subsection line, that is sufficient.

The need for descriptive ties to the adjoining land is paramount in describing those lines which are not established by rules or not shown on previous maps and for which there is no other record information.

Section land development applies to what is known as the PUBLIC DOMAIN area of the United States of America (including Alaska) as delineated in Figure 3-7a. This was divided into various sized areas each of which is governed by a Principle Meridian as shown in Fig. 3-7b. All came under the Rules of Survey of the Public Lands except those lands in private grants in the midwest such as the Virginia Military Reserve, the Greely French Claims, the Connecticut Western Reserve etc and those Ranchos in the southwest given by the prior Spanish or Mexican rulers which were to be honored according to the Treaty of Guadelupe Hildago. Within these sections, numerous metes and bounds descriptions have been created with ties and controls to original section and sub-section lines and corners.

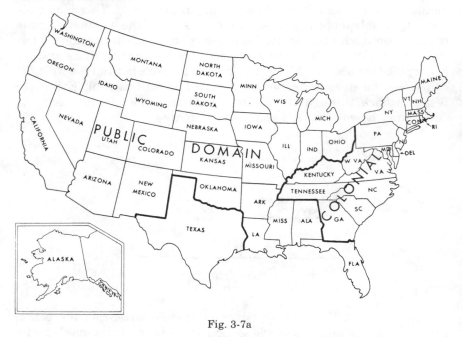

Fig. 3-7a

Texas developed its own type of subdivision through grants of leagues which were called by the name of the grantee such as the "Nash League"

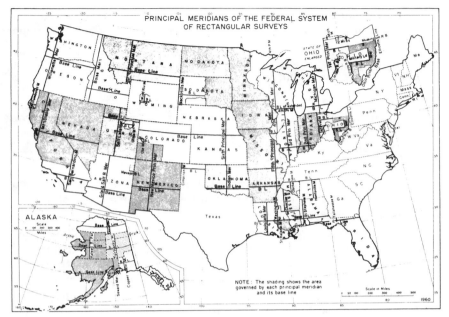

Fig. 3-7b

the "Lorenzo de Zavala 1/3 League." It also has some sectionized lands developed by its own surveyors. Many metes and bounds descriptions exist as a result of cutting up these "master areas."

The area which includes the original colonies and Kentucky and Tennessee was not sectionized. All of the original parcels of land were described by metes and bounds. In later years, subdivision maps were developed and used for the conveyance of parcels shown thereon.

Metes and Bounds

"Metes" refers to the bearings and distances cited as courses and "Bounds" refers to the monuments which define the boundary or limit of the property. When you are given a metes and bounds description from which you are to draw a map, there is a definite procedure to follow.

Reading through the description, make notes of the following items:

1. City, _____;
2. County (Borough or Parish), _____;
3. State, _____;
4. Is the land all or part of a lot or parcel shown on a map? If so, what is the lot or parcel number, letter or name, _____?
5. What is the name or number of the map, _____?

6. If the map is officially filed or recorded, what is the correct reference file no. _____or book _____ and page _____?

7. If the land is part of the Public Lands (that is section land), what is the section number _____, Township _____, Range _____ and Meridian _____?

8. If the land is part of a private land grant (such as the H. Lewis 752 ac Tract or the Rancho Pascuala), what is the correct reference to its map or its creation by book _____, page _____?

The above items are necessary first for the information to be shown in the caption on your map and second for the source of control from which you will develop and draw your map.

After determining the relationship between the POINT OF BEGINNING in your description and your points of control on the map of which you are a part, make a *rough* sketch of the description to see the general shape and size of it so you can estimate by scale relationship the size of paper you will need. If you are limited to a certain size map, then you will need to determine the scale which will accommodate the space that is available. *Be sure you have enough room to also show any external points of control.*

Next, establish a base line for the bearings however it may be delineated in the description. If you have a part of a lot or parcel on a recorded or filed map, then presumably you will use the same base of bearings as is on that map. If, however, the first line of your description reads, "thence along the westerly line of Jones' land, N 02° 38' E . . . . .," then that is your base of bearing with which you begin your map.

There are three methods for laying out and developing a map: —

1) Having established your controls and base of bearing line, proceed to plot each successive line or curve in the description. Here is a simple one for illustration:

That portion of Lot 9 in the RAY TRACT in the County of Vale, State of Wildhorse, as per map recorded in Book 10, Page 5, of Maps in the office of the County Recorder of said County described as follows:

Beginning at the southwest corner of said Lot 9; thence along the westerly line thereof N 0° 03' 20" W 15.00 feet; thence parallel with the southerly line of said Lot N 89° 56' 40" E 100.00 feet; thence N 44° 56' 40" E 56.57 feet; thence N 89° 56' 40" E 34.25 feet; thence S 30° 03' 20" E 63.51 feet to the southeast corner of said Lot; thence S 89° 56' 40" W 206.00 feet to the point of beginning.

Having copied Lot 9 and a portion of the street in front of it from the map of the RAY TRACT, you are then ready to cut-in the metes and bounds description. It states that the westerly line of the lot (which is obviously also the easterly line of the street) bears N 0° 03′ 20″ W, therefore it is the base of bearing for this description. Your first course is established by measuring 15.00 feet along said westerly line northerly from the southwest corner of the Lot. See Figure 3-8.

Map of Survey
of
a portion of LOT 9 in the RAY TRACT
County of Vale, State of Wildhorse,
as per map recorded in Book 10, Page 5, of Maps,
records of said county.

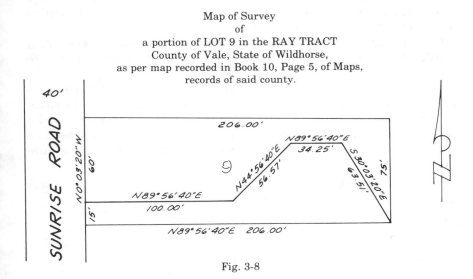

Fig. 3-8

Referring back to Figure 2-5 and the explanation for deriving angles from bearings, the angle between the west line of this lot (N 0° 03′ 20″ W) and the next bearing (N 89° 56′ 40″ E) is the sum of the two bearings (since they are both in the north quadrants) which equals 90° 00′ 00″; therefore, you draw a line at right angles from Sunrise Road into the lot and scale 100.00 feet on it.

The bearing of the next line is in the same quadrant as the previous one so the difference between them (N 89° 56′ 40″ E − N 44° 56′ 40″ E) equals 45° 00′ 00″ and it is the deflection angle northerly from the easterly prolongation of the last line. Set this off the protractor and mark the distance of 56.57 feet along the line.

Since the bearing of the next line is the same as the first line into the lot, it is only necessary to draw a line parallel with that first line easterly from the last point and mark the distance of 34.25 feet on it.

Referring again to the explanation of Figure 2-5 for the next step, it is necessary to reverse the direction of the bearing N 89° 56′ 40″ E to S 89° 56′ 40″ W and add it to the bearing of the next line S 30° 03′ 20″ E (since now they are both in the south quadrants) which gives 120° 00′ 00″ for an interior angle. The line drawn along this angle and the distance of 63.51 feet should go exactly to the southeast corner of the lot; if it does not, you will make it go to the corner because the tie in the description is to the lot corner.

The angles derived from the bearings in this description are for the use of a protractor in laying out the lines. If a drafting machine is available, then the lines can be drawn directly from setting the bearings on the machine and this is the fastest manual method of drawing a map from a metes and bounds description.

2) If you were to run a traverse of this description and develop a set of coordinates, you could plot it in the same manner as shown in Figure 2-7.

3) Still another method of plotting is by latitudes and departures which shortcuts and deletes coordinates as such but follows the same form of development. Without reciting the metes and bounds description, let us take a traverse of such a boundary and lay out north-south distances and the east-west distances in consecutive order. See Figure 3-9.

Course	Distance	Lat. North	Lat. South	Dep. East	Dep. West
N77° 00′ 20″ E	133.42′	30.00		130.00	
S18° 26′ 06″ E	94.87′		90.00	30.00	
S79° 41′ 43″ W	111.80′		20.00		110.00
N32° 00′ 19″ W	94.34′	80.00			50.00
		110.00	110.00	160.00	160.00

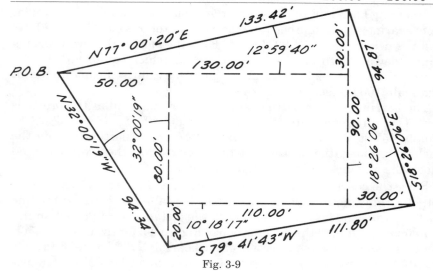

Fig. 3-9

Starting at the point of beginning (POB), the latitudes will be marked on north-south lines and the departures will be measured at right angles thereto on east-west lines. The reason for using this method is that it can be done without a protractor, although it would be desirable to check the resultant lines for the bearings and distances cited in the description. Of course here too, coordinates could be developed and applied as a check.

EXERCISE 3-2: Three methods have been explained for the transposing of a metes and bounds description into a map. Take the following description and draw a map of it using the first method.

Facts: The southwesterly line of this Lot 15 is 235.00 feet (77.10m) long and is also on the northeasterly line of Wood Street and bears S 68° 00′ E. The northwesterly line of this lot is 184.00 feet (60.37m) long and bears N 35° 00′ E. The southeasterly line of the lot is 216.00 feet (70.87m) long and bears S 20° 00′ W. The lot will draft onto a letterhead size sheet using either a 1″ = 30′ scale or a 1:125 scale.

Description:

That portion of Lot 15 of Fred Macrae's Tract in Calcasieu Parish, Louisiana, as per map filed in the office of Public Records described by the following courses:

1 — Beginning at a point on the northeast line of Wood Street S 68° 00′ E 22.40 feet (7.35 m) from the most westerly corner of said Lot,

2 — N 63° 20′ E 38.50 feet (12.63 m),

3 — N 24° 30′ E 47.42 feet (15.56 m),

4 — S 87° 25′ E 87.86 feet (28.83 m),

5 — N 19° 20′ E 35.64 feet (11.69 m),

6 — S 68° 00′ E 78.83 feet (25.86 m),

7 — S 24° 00′ W 141.15 (46.31 m) to Wood Street,

8 — N 68° 00′ W 182.62 feet (59.91 m) along Wood Street to the point of beginning.

* * *

When a course ties to a point, a line, a monument, etc., that thing is the controlling factor and if the tie is to a line only, then the distance must give, while if the tie is to a point, then both the distance and bearing in the line before it must give to the position of that point.

If courses 6 and 7 were changed to read:

6 — S 68° 00′ E 78.83 feet (25.86 m) to a point on a line parallel with and 29.96 feet (9.83 m) northwesterly from the southeasterly line of said Lot which point is N 20° 00′ E 138.00 feet (45.28 m) from the northeasterly line of Wood Street,

7 — S 20° 00′ W 138.00 feet (45.28 m) to Wood Street,
draw the additional lines necessary to show their position with heavy
dash lines.

If either or both the drawing and mathematical traverse do not close, a
quick way to check them is to draft vertical and horizontal lines from the
beginning of each course to its end similar to Figures 2-7 and 3-9.

*   *   *

Coordinate Relationships.

We have discussed the use of coordinates for laying out traverses but
let us consider their value from a legal standpoint and then their
application to maps.

Land titles stand on the framework of legal descriptions developed
from, and their continuity supported by, surveys. Surveying relies on
the science of mathematics in the form of measurements of distances
and angles for input information. The determination of survey lines and
points representing ownership however is controlled by legal doctrines
concerning boundaries, monuments, topography, priority in time and
facets thereof. This is why the professional surveyor must be well versed
in both the discipline of measurements *and* the knowledge of legal
principles involved in the perfection of title to land.

Coordinates are an excellent tool for many purposes and they assist
the surveyor in numerous ways, both in the field and office. It is possible
to devolve coordinates from a triangulation network down to any point
desired.

Some states have provided for the acceptability of coordinates assigned to parcel corners, others have merely legislated the creation and
development of a state plane coordinate system, while a few have been
specific in recognizing coordinates and establishing a statewide system
but denying any legal status of coordinates over and above ties to
matters of record insofar as they may control the boundaries of ownership.

This does not mean you cannot have descriptions with coordinates but
from the standpoint of title, any such application should be on a parenthetical basis so as to retain the certainty of legal control. Again,
consult *your* state codes to determine the extent to which coordinates
have been established, the form of designation assigned to them and
their legal connotation.

The assignment of coordinates to a point or monument by a surveyor,
a photogrammetrist or a draftsman merely identifies an accurate
method for reestablishing that point on the ground or on paper; it does
not make a determination of its title value. Until or unless the coordinates of such a point are tied to, or made a part of, matters of record in a
description, their value is no greater than any other dimension.

It has also been suggested that all monuments be disregarded and that coordinates be the sole label and responsibility of property corner designations. Many decisions by the courts in land boundary cases are based on surveys and those surveys in turn are based on monuments and metes and bounds descriptions. So, regardless of anyone's attitude toward monuments, it is necessary to live with them where necessary as embraced by the law until it is changed. This is why surveyors, or anyone else, cannot ignore *properly* established monuments. Even if the law were changed to disregard all monuments, it is extremely doubtful that it would be retroactive because it would cause too many economic problems. Therefore, all those previously acknowledged would still have to be honored. Furthermore, many people would become very skeptical of coordinates *they* could not understand in lieu of the physical monuments they could see.

Another objection to the sole use of coordinates is that one or more errors in the figures can throw the entire area out of title and you have no warning of it. In other words, a miscopy or transposition of a number could be made accidentally and there would be no signal to inform anyone of the mistake until extrinsic conditions brought it to light.

With legal descriptions, you have built-in checks with ties and monuments to which lines must conform. Properly written descriptions are inherently correct insofar as legal title is concerned. The legally allowable variation in measurement by ties permits the survey and ground conditions to be brought into agreements with each other and with title.

In a description incorporating coordinates, the fundamental specifications are that they refer to the correct zone, if there is more than one, the state designated nomenclature of the Coordinate System and if it is Grid or some other type of reference.

Although a single coordinated point in a mathematically closed description would position the property (provided the bearings were on the same basis as the coordinates used), it is preferable that there be at least two coordinated points. It is not necessary to assign coordinates to every point inasmuch as they can be derived from the traverse; rather, it would be advisable to choose pertinent control points that have been tied by actual survey to, or can be accurately computed from, higher order controls. The following description illustrates the application of coordinate inclusion.

EXAMPLE:

That portion of the SE 1/4 of the SE 1/4 of Section 15, T ___ S, R ___ E, of the _____ Meridian in the County of _____, State of _____, according to the Official Plat of said land filed in the District Land Office described as follows:

Beginning at a nail set in a square cut stone on the south line of said Section in the middle of Three-mile Road being the southwest corner of the land described in deed to Su San Kim recorded in Book _____, Page _____, of Deeds of said County at a point N 89° 48′ 16″ W 400.00 feet from the southeast corner of said Section, being a marked stone having established grid coordinates of (X) _____, (Y) _____, of Zone 2 of _____State Coordinate System; thence along the westerly line of said Kim land (bearings based on grid meridian of said Zone 2) N 28° 17′ 32″ E 200.00 feet; thence N 66° 32′ 51″ W 331.64 feet to a point on the east line of the WW Tract as per map recorded in Book _____, Page _____, of Maps in the office of the County Recorder of said County distant along said line S 0° 10′ 24″ E (record bearing of said east Tract line shown as SOUTH) 102.68 feet from a 2″ (O.D.) iron pipe at an angle point in said Tract line having grid coordinates of (X) _____, (Y) _____, of said Zone 2; thence along said Tract line S 0° 10′ 24″ E to said south line of Section 15; thence S 89° 48′ 16″ E to the point of beginning.

When you encounter descriptions with coordinates from which a map must be developed, there is one item in particular that requires special consideration; it is the relationship between the ground measurements and the coordinates which are usually State Plane Coordinates. In other words, you want to be aware of the fact that those coordinates are based on a sea level surface of the earth which for practical purposes is considered to be an ellipsoid. If your job, for example, is a mile above sea level, the geodetic distance between two points computed from true coordinates will not be the same as the ground level distance between those two designated points as illustrated in Figure 3-10. It is not the objective here to go into the mathematics of converting geodetic lengths to ground level distances nor vice versa but to alert you to the existence of such a condition. The methods and calculations for such operations are available in manuals prepared by the U.S. Department of Commerce, Coast & Geodetic Survey, now National Ocean Survey

Ground level

Sea level

Center of Earth
Reduction to sea level
Fig. 3-10

of National Oceanic and Atmospheric Administration of the U.S. Department of Commerce, and in texts.

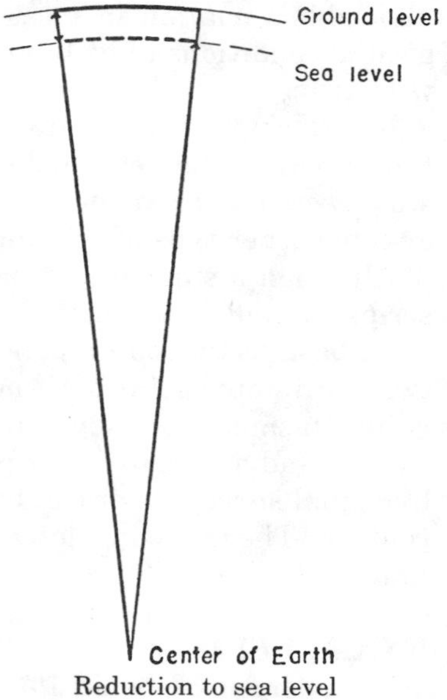

Sometimes coordinates are assigned to the adjusted field work without going to sea level. Be sure you determine if this has been done and, if so, what you are supposed to use. Perhaps an example of what happened in one case will best illustrate the caution necessary to be exercised. A field party was sent out to set corners shown on the map given to them by the office calculator. After setting a group of points, the distance to a tie point previously established did not check by several tenths of a foot which was too much. Questioning the office revealed the fact that the calculator had not applied the Scale Factor (sometimes called the Constant [K] for Conversion or Combination Factor). This notice of the Scale Factor and whether the measurements shown on the map are related to the grid or ground basis is very important.

There are two types of grids. When the State coordinate system is based on the Lambert projection, it is designated as being on the *Lambert grid* and if based on the transverse Mercator projection, it is called the *transverse Mercator grid.*

There is another use of coordinates which must not be confused with any assignment of them to a legal corner, in other words, this is a situation whereby they are used *solely* as a tool to accomplish the end, not as a label on the true corner.

Because of physical obstructions that prevent running a straight line in the field, such as trees, rocks or even buildings, the survey crew may have established certain points on a traverse line along the way but could not locate the property corner; therefore, they ask the office to develope an angle and distance tie from their nearest set point to that corner so they can find it. You then research the necessary information to determine where that corner *should* be and you develope coordinates, either real or assumed, and relate it by a forced closure to their field point. From this information, they can measure in and look for the corner and when they find it, make ties which will give you the actual position to show on your map.

Right-of-way or Easement Strips

The following will be a brief review of the status and meaning of easements because you need to know what they represent and their relationship to each other and to the whole area when you draw them on a map.

While the absolute ownership of land is expressed as a *fee simple absolute,* of which the short form "fee" implies the whole of the meaning, there are various forms of rights, or interests or claims that may be placed upon the land, thereby encumbering the land. These may take the form of easements, rights-of-ways, monetary interests, water rights, mineral interests and various forms of rights for individual benefits. It is not the intent here to explain all the ramifications of these encumbrances but to alert you to their existence and use.

An EASEMENT, usually for the benefit of one or more individuals, is an interest in the land of someone else and therefore constitutes an encumbrance on another's land. A fee owner cannot own an easement over his own land. An easement carries a limited and non-possessory use by citation. It is created by grant or agreement, expressed or implied. It is transferred subject to the rules of real property. The easement holder has only such control as may enable him to use it as specified in the grant thereof. Ordinarily, this does not exclude others from making use of the same land in a way which does not interfere with the enjoyment of prior rights. In other words, there may be multiple easements over the same strip or area of land but each one must honor those whose rights were acquired previously. An easement granted through the medium of a document is never lost as a result of non-use.

Land is held as one of two estates in respect to easements: as a *dominant tenement* or as a *servient tenement*. The *dominant tenement* is the land to which an easement is attached or the service is owed. The *servient tenement* is the land which is burdened with a servitude. When one has an easement over adjoining land, it benefits his fee ownership and his land becomes the dominant tenement but the same easement is an encumbrance on the adjoining land which thereby becomes the servient tenement. See Figure 3-11.

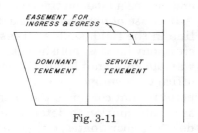

Fig. 3-11

There are two classifications of easements, *appurtenant* and *gross*. An *appurtenant* easement is created, attached to and for the benefit of the land of the owner of a dominant tenement. The easement for ingress and egress shown in Figure 3-11 is known as an appurtenant easement for the benefit of and attached to the dominant tenement. An appurtenant easement once established may pass without a specific grant.

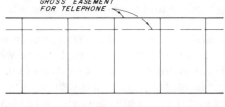

Fig. 3-12

A *gross* easement is a right in another's land, which right is not for the benefit of any land owned by the easement holder. "An easement for telephone purposes over the North 6 feet of Lots . . . . .," would be a gross easement as shown in Figure 3-12. An easement in gross is not attached to the land, as is an easement appurtenant, but is considered to be a personal or corporate right attached to the easement holder, yet that right does constitute an interest in another's land, which again is the servient tenement.

Your services may be requested to graphically position an easement in relation to the boundaries of either/or both the dominant tenement and/or the servient tenement.

In reading easement descriptions, it is important to note the *purposes and uses* of the easement which are expressed because by rule, it is a right to a limited use or enjoyment of land. For example, if one is acquired for "ingress and egress" but subsequently it is desired to install a water line and power lines, by the qualification attached to it there was no provision made for the latter uses so, consequently, the granted easement is insufficient. Furthermore, it is desirable to have the easement designate if it is appurtenant, and if so, to what property.

You will read many documents wherein the purpose(s) of the easement is clearly set forth, such as:

"A non-exclusive easement for road purposes . . . . ."
"An easement 10 feet wide for the installation and maintenance of two oil pipe lines . . . . ."
"An exclusive easement for ingress and egress and utilities in and over a strip of land 14 feet in width . . . . ."
"A one rod wide easement for a water line and driveway for maintenance . . . . ." etc., etc.

This is good because it sets forth its operative use.

Overburdening

A word about overburdening an easement. In a simple subdivision of land as illustrated in Figure 3-13, the use of easements over Parcels V and X by Parcel U is normally acceptable, but if the owner of Parcel U divides it into ten parcels and grants each of them easements over V and X, it would be considered an overburden on the easements over V and X and probably held as unreasonable. The title to the easements held by the owners in subdivided Parcel U is good; it would be a matter of unauthorized increase in burden, opening the door to an injunction against the increased use. There is one case (Sec. 28, C.J.S. 62) where an easement was regarded terminated by the increased burden because, under the circumstances, it was impossible to separate the authorized and unauthorized uses.*

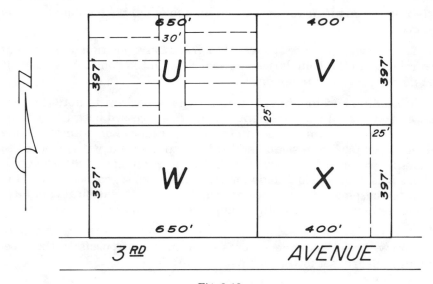

Fig. 3-13

Rights-of-Ways

Although a right-of-way is technically an easement, it is peculiar in that it is expressly for "passage" purposes; it may be for a railroad, pipe lines, pedestrians, vehicles, a canal, etc. In essence, a grant or creation (possibly without a grant) of a right-of-way establishes a privilege to pass over another's land, under in the case of a tunnel and aerially over in the case of a bridge. It may be for the benefit of an individual, a group or class of persons or the public. Because *right-of-way* is ofttimes used to describe the strip of land itself, questions arise for judicial determination as to whether that strip is in fact an easement, or a fee, subject to the right of a specific use. Another definition for right-of-way is a composite of strips of land over which an easement passes.

*PRIVATE EASEMENTS from Proceedings of California Land Title Association, May 1951.

Strip Easements

As the name implies, a strip of land is created either en toto or in segments, either in the status of an easement or a fee limited to certain rights as stated above to accommodate a certain use.

The strip "en toto" description is easy to create because after making a tie, at one end, it takes off across country with little or no concern for intermediate ties to title lines until it ends somewhere. Problems sometimes arise when one or more owners along the way want finite determination of their boundary line in relationship to the easement *prior* to conveyance. At other times problems arise when it is found later by field location that the right of way in fact cuts across corners of other land for which conveyances were not acquired.

One such case would be a strip passing through a parcel marked 20 acres on a map, as shown in Figure 3-14 where one owner had acquired the North 12 acres and later the other owner had acquired the South 8 acres, but actually there was an excess of about 2/3 of an acre which belonged to neither one of these owners by conveyance and could not be prorated. This would leave a sliver between, as shown in Figure 3-15,

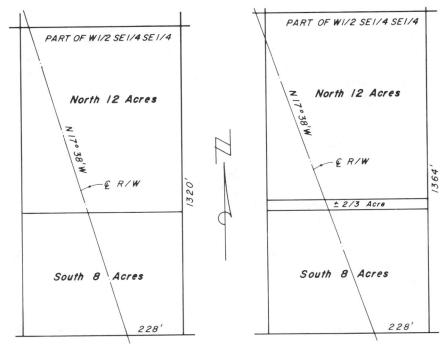

Fig. 3-14   Fig. 3-15

possibly in the original owner of the whole parcel, possibly in the status of after-acquired title by the buyer of the South 8 acres, or, under prescriptive right by use of either/or both owners and claimed by either/or both sides, in which case it would then require a quiet title action. Consider what would happen to this strip easement if it were on the diagonal through this parcel with a tie only at one end. The description of the center line of this easement based on record might read:

> "Thence from a point on the south line of said west half of the southeast quarter of Section 26 distant westerly 228 feet from the southeast corner of said west half, North 17° 38′ West 1385.1 feet to the north line thereof."

However, with the position of the north line of this parcel actually 1364 feet instead of 1320 feet north of the south line of the section, the line of right of way which must hold the bearing (since there was no lateral tie given on the north line) finds itself in foreign property on the west, thus requiring an additional conveyance or a revision of the right of way. Sufficient preliminary field survey work would reveal such a problem and save subsequent embarrassment.

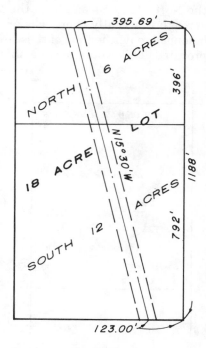

Fig. 3-16

Another similar example is shown in Figure 3-16 but with a different problem which results from there being a tie along the north line as well as along the south line. According to information obtained from the record, the scrivener began the center line of the easement "at a point on the south line of said lot, 123 feet West of the southeast corner; thence North 15° 30′ 00″ West to the north line of the South 12 acres of said lot," for the southerly parcel; and began "at a point on the south line of said lot, 123 feet West of the southeast corner; thence North 15° 30′ 00″ West to the north line of the South 12 acres of said lot and the true point of beginning; thence North 15° 30′ 00″ West to a point on the north line of said lot 395.69 feet West of the northeast corner thereof" for the northerly parcel. Insofar as the *record* was concerned, the information conformed;

however, by inspection of Figure 3-17 showing the ground measurements, it is clear that the above form of description would necessitate an angle at the north line of the south 12 acres in order to meet the tie on the north line of the lot, and it would also traverse the sliver between the two parcels.

If, on the other hand, the scrivener had used the reverse order for the northerly parcel and said, "beginning at a point on the north line of said lot, 395.69 feet West of the northeast corner thereof; thence South 15° 30′ 00″ East to the southerly line of the North 6 acres," he not only would leave a gap but even the prolongations of each strip would be offset one from the other.

Neither could the scrivener use the inclusive form of that portion of the South 12 acres included within the following described strip, and

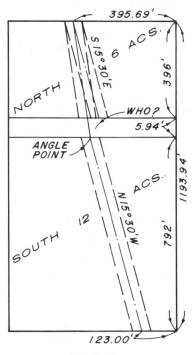

Fig. 3-17

that portion of the North 6 acres, etc., etc., and use the center line based on the ties by record because the prolongation of the lot as revealed by measurements would cause an uncalled-for angle in the line and traversing over an intervening segment of land, the ownership of which was not known. So, which line should be used, and when is it possible, or is it possible, to use record information to accomplish the goal when actual measurements to the contrary are not minor but major differences from the record?

In this case at hand, each owner had bought his respective number of acres and had lived in the firm belief that that was exactly what he had, and each having planted oranges and each having used the open area along their adjoining boundary as a common turn-around for their cultivation equipment, they therefore had no finite line of demarcation between them.

To understand the situation clearly, there should be full comprehension of the relationship between the following four elements:

1. record information from the document,
2. the strict survey of the record per se,
3. the relationship of monuments, and
4. the juxtaposition of any physical occupation.

Although it is intended that these items coincide, many times they do not.

The map that you draft should reflect these conditions, not only for the benefit of the right of way man, but also for the grantor involved and the contractor doing the improvement work. One unaccustomed to survey description work could easily be unaware of the fact that a monument may not, in truth, be the corner of his property.

When the office man is planning the right of way and he finds that the reconnaissance surveyor shows in his field notes distances, angles and monuments at variance with record information, the creator of the easement-to-be needs to superimpose one on the other and then analyze the findings to determine (1) what *must* hold, (2) what *may* give, and (3) what adjustments may be required. It may even be necessary to get additional field or record information or both in order to make the decision.

Matters which are already on record must hold in their proper delineation. The lines which may give are controlled by ties or calls to other things.

The metes and bounds description for individual parcels with an accompanying map is desirable because it pinpoints the area to be acquired from each grantor and makes it easier for him to comprehend. For this, the preliminary survey work must be sufficiently comprehensive to enable the draftsman to properly develop the map.

Descriptive Forms.

The most common form of describing a strip of land is by use of a center line control such as:

> "A strip of land 50 feet wide, the center line of which is described as follows:"

or

> "An easement for road purposes 40 feet wide lying 20 feet on each side of the following described line:"

Either because of physical circumstances or by lines of ownership, a line, other than the center line, sometimes called the survey line, may be used to control the strip such as:

> "A strip of land 75 feet wide lying 25 feet north and 50 feet south of the following described line:"

or if the strip follows around a hill or for some other reason changes direction, the wording might be thus:

"A strip of land 75 feet wide lying 25 feet northerly, northwesterly and westerly and 50 feet southerly, southeasterly and easterly of the following described line:"

A third way is by a line completely offset on one side of the strip using the form:

"A strip of land 46 feet wide, the southwesterly and southerly line of which is parallel and concentric with and 15 feet northeasterly and northerly of the following described line:"

There are situations where the terrain is such that extra widths are needed to accommodate cuts and fills on slope land. Very often, in addition, areas of various sizes are needed as temporary easements for construction. The drawing in Figure 3-18 shows such a combination of

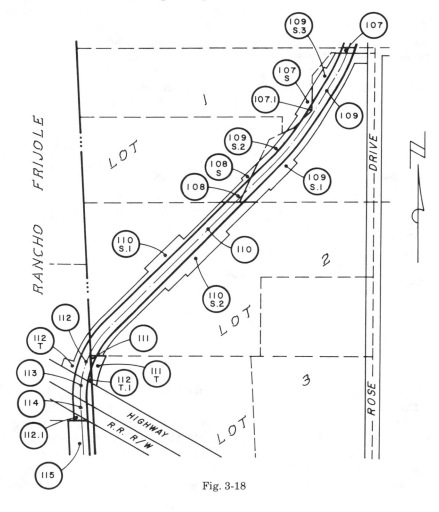

Fig. 3-18

permanent, slope and temporary easements divided among their respective ownership parcels.

You may find additional segments with parcels of land under single ownership using one of the strip type descriptions mentioned above followed by a form such as:

"Together with a temporary easement for construction purposes over a strip of land 40 feet wide, the northerly line of which is the southerly line of the hereinabove described strip bounded on the west by the radial line passing through the easterly end of that curve cited as having a radius of 800 feet and length of 367.43 feet and bounded on the east by a line which bears S 2° 41' 32" W from the easterly terminus of the above-described center line."

Additional areas for slope easements can be handled in the same way if the required shape will permit, but sometimes the configuration is such that an entirely distinct and separate metes and bounds description is necessary. The numbers shown represent parcels from different owners for the permanent easement. The addition of the letter "S" indicates an added area for slope easement and the addition of the letter "T" indicates those areas needed temporarily for construction purposes.

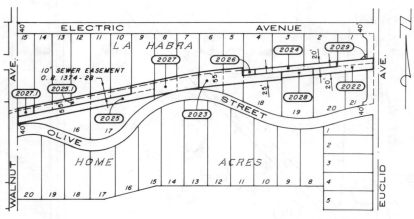

Fig. 3-19

Another example of parcel taking is illustrated in Figure 3-19 showing the parcel for temporary use designated with a decimal (2025.1). The descriptions for most of these parcels are simply "The southerly _____ feet of Lot Number _____, etc . . . . . ." or "The northerly _____ feet of Lot Number _____, etc. . . . . ." Of course, the irregular shape taken in Lot 1 is described by metes and bounds. Considering the chronological order of procedures in acquiring R/W parcels, it is not always feasible to prepare a complete map ahead of time; however, the use of easement parcel maps "attached hereto and made a part hereof" with each document of conveyance not only illustrates the description but it assists the negotiator in his dealings with the owner and makes it easier for others who subsequently have occasion to use such a document.

Stationing

Some survey work is developed on a station-to-station basis; that is, the line (center line of R/W, center line of construction, arbitrary survey line, property offset line or whatever) is measured from a 0 + 00 point of beginning and everything along the way such as ditches, pole lines, pipe lines, culverts, beginning and end of curves, railroads, streets, etc., etc. is assigned a station according to its horizontal linear measurement from 0 + 00. Obversely, any distance between objects is equal to the algebraic difference between their respective stations.

When this type of line intersects another line with stationing, an equation is expressed at that point relating one to the other. If there is partial realignment, for instance, in the size of a curve which changes the position of the BC and EC, then an equation must be established at one or both of those points in order to maintain a mathematical relationship. It is very necessary to have these points of equation accurate and correctly placed on the map or construction plans.

Multiple Parcels

On occasion, there may be a situation which requires that branch strip easements be attached to the main easement. Generally speaking, multiple parcels under one caption are not recommended, but if *all* of the area encumbered by the description is included in the caption *and* the respective pieces are properly tied to the main strip, then it may pass scrutiny well enough to be acceptable. The following is an example: (See Figure 3-20)

> An easement 20 feet wide over that portion of Lot C in Tract No. 436 in the County of Missoula, State of Arizona as per map recorded in Book 8, Page 4, of Maps in the office of the County Recorder of said County and those portions of Farm Lots 7 and 8 in the Butler Tract in said County as per map recorded in Book 2, Page 13, of said Maps and that portion of the southwest quarter of Section 10, T 2 N, R 4 W of the Gila and Salt River Meridian in said County, according to the official plat of said land filed in the District Land Office, the center line of which is described as follows:
>
> Beginning at the west quarter corner of said Section 10, being the original stone marked 1/4; thence S 00° 54′ W 389.42 feet on the west line of said section, being also the center line of Taco Road;

thence S 89° 06′ E 20.00 feet to the True Point of Beginning on the east line of said Road; thence S 51° 17′ E 364.00 feet to Point A; thence S 51° 17′ E 189.68 feet; thence S 89° 06′ E 158.00 feet to Point B; thence S 89° 06′ E 147.00 feet to Point C; thence S 89° 06′ E 206.00 feet; thence N 58° 32′ E 458.62 feet, together with:

Parcel A: A 10 foot easement the center line of which begins at the hereinabove described Point A; thence S 89° 06′ E 289.00 feet.

Parcel B: A 6 foot easement the center line of which begins at the hereinabove described Point B; thence S 00° 54′ 100.00 feet.

Parcel C: An 8 foot easement the center line of which begins at the hereinabove described Point C; thence N 00° 54′ E 109.00 feet.

The side lines of said 20 foot easement to be extended or shortened to meet at angle points and to terminate at the said ease line of Taco Road.

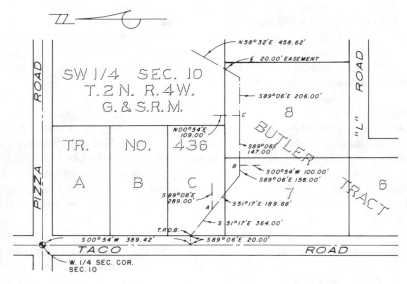

Fig. 3-20

Horizontal Curves in Surveys and Descriptions

Curves become an integral part of highways, railroads, canals, right-of-ways, subdivisions, and property boundaries when the surface conditions of the land or the esthetic design of subdivisions require such configurations; therefore, one must know the methods of creation of curves, their interpretation and proper delineation for correct descriptive application. In this discussion, curves will be treated only in the horizontal plane.

A CURVE is a segment of the circumference of a circle. Its end limits are the radii of the circle and the angle between those two radii is the central angle or delta of the curve.

The correlated parts, known as elements of a curve, are: radius (R), length (L), central angle or delta (Δ), semi-tangent (usually just called tangent) (T), and chord (C). See Figure 3.21.

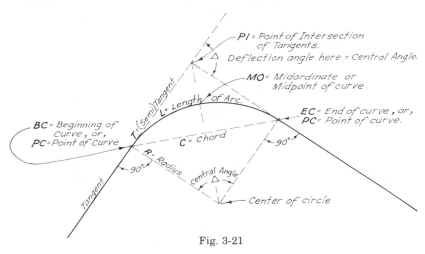

Fig. 3-21

-a- RADIUS is the line between the center of the circle and its circumference.

-b- The LENGTH of arc is the distance along the curve.

-c- The CENTRAL ANGLE or DELTA is the angle subtended by the radii.

-d- The SEMI-TANGENT (generally just called TANGENT) is the extension of the straight line (sometimes called the tangent line) which precedes or follows the curve to its intersection with the other semi-tangent.

-e- The CHORD is that segment of a straight line which is intersected by two points on a curve, or, in other words, the straight line distance between the two ends of a segment of arc.

The illustration in Figure 3-22 shows the application of the following qualifications of curves:

-a- A *simple curve* is a single segment of arc.

-b- *Compound curves* are a group of two or more segments of arcs having different lengths of radii on the same side of the curves, common radial lines and common tangents.

-c- A *spiral curve* is a collective group of multiple compound curves having radii of successively decreasing or increasing lengths.

-d- A *reverse curve* is a segment of arc of which the center of its circle is on the opposite side of the adjoining curve and of which one of its radial lines is the prolongation of a radial line of the opposite curve.

-e- *Tangent curves* have a common radius or a prolongation of each other's radius (though they may be of different lengths); therefore, compound curves and reverse curves are by definition tangent.

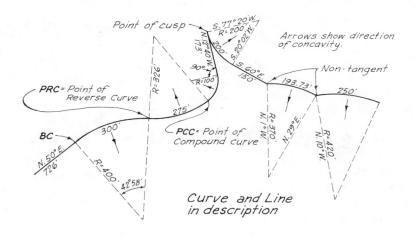

Fig. 3-22

-f- *Tangency* of a curve to a line or curve means that the radius line is perpendicular to that line or the tangent of the curve at that point of curve.

-g- *Concavity* relates to the enclosed side of a curve toward the center of the circle. The general direction given for the concavity is that of the radial line from the midpoint of the arc toward the center.

-h- *Convexity* (seldom used) relates to the outside face of a curve away from the center.

-i- *Non-tangent* relates to the fact that the point of curve is not tangent to the preceding line or curve. This is also known as a *broken-back* curve.

-j- A *cusp* is the meeting of two curves or a curve and a straight line having the same general direction. The point of a crescent is a cusp.

-k- The direction given along the curve is a general one (NE'ly, S'ly, etc.) for the total arc; however, if the total angle is more than about 120°, two or more directions may be combined, such as W'ly and NW'ly. See Figure 3-23.

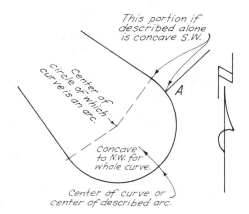

Fig. 3-23

In the case of a cul-de-sac as in Figure 3-24, three directions would be desirable, i.e., "thence 286.74 feet north-westerly, easterly and southwesterly, along a curve concave to the south having a radius of 47.00 feet."

Other points related to the curve shown in Figure 3-21 and used in descriptions are:

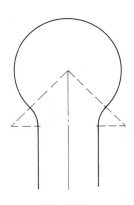

Fig. 3-24

-a- Beginning of curve (BC) which is the point in the description where the curve begins regardless of whether it is tangent or non-tangent. It is sometimes called a point of curve (PC).

-b- Beginning of compound curve (BCC) which is the point in the description where one curve ends and an adjoining tangent curve having a common radius begins. It is sometimes called a point of compound curve (PCC).

-c- Beginning of reverse curve (BRC) which is the point in the description where one curve ends and an adjoining tangent curve concave in the opposite direction begins. It is sometimes called a point of reverse curve (PRC).

-d- Midordinate (MO), or midpoint of curve, which is the halfway point along the arc between the beginning and end of the curve. It lies on the bisection of the central angle. This is very seldom used.

-e- Point of intersection of tangents (PI) is the point at which the prolongation of the tangents of a curve, one passing through each end of the curve, intersect, but not possible when Δ is more than 179° 59' 59".

-f- End of curve (EC) which is the point in the description where the curve ends. It is also sometimes called a point of curve (PC). This statement is seldom used.

-g- The center of a circle is the point from which the radial line bears.

All of the mathematical data of a curve can be derived from the facts given for any two elements. The most commonly used ones are the radius and length. Sometimes the central angle is added. In some instances, the bearing and length of the chord is given. The third element can be used for a check on the others unless it creates conflict, in which case, extrinsic evidence may be necessary for a solution.

Before proceeding with the multiple curve calls, let us consider the problems involved by the use of the words "right" or "left". The use of the words "right" or "left" to connote one side or the other is not fully satisfactory. In the early days, the courses of a river would be described as following the left or right bank going downstream. Attempts were made to adapt this to other uses, but inherent variables made it uncertain. See Figure 3-25. A compass direction is not the best either; a positive bearing is much to be preferred.

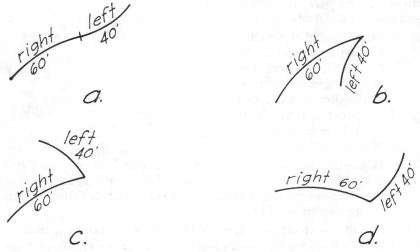

Fig. 3-25

If your description reads, "Thence along a curve to the right 60 feet, thence along a curve to the left 40 feet . . . .," then examples a, b, c and d in Figure 3-25 would all be apropos but certainly not finite to the particular condition at hand.

Consider a more certain way by the use of compass directions, still referring to Figure 3-25.

    a. "Thence northeasterly 60 feet along a curve concave to the southeast having a radius of 100 feet to the beginning of a reverse curve concave to the northwest having a radius of 60 feet; thence northeasterly 40 feet along said curve."

Unless otherwise stated, both a reverse and a compound curve are tangent to their previous curves so this part is compatible with Figure 3-25a.

Continuing with examples b, c, and d:

    b. "Thence easterly 60 feet along a curve concave to the south having a radius of 60 feet to the cusp of a curve concave to the southeast having a radius of 50 feet; thence southwesterly 40 feet along said curve."

And: c. "Thence northeasterly 60 feet along a curve concave to the southeast having a radius of 200 feet to the cusp of a curve concave to the southwest having a radius of 70 feet; thence northwesterly 40 feet along said curve."

Also: d. "Thence southeasterly 60 feet along a curve concave to the south having a radius of 200 feet to the beginning of a non-tangent curve concave to the northwest having a radius of 75 feet; thence northeasterly 40 feet along said curve."

In each of these described examples, there is no question about the *general* position of each of the curves, their relation to each other, and their direction of travel. However, this is still not finite enough for surveying because each of the eight general compass directions — north, northeast, east, southeast, etc. — include a spread of 45° or 22° 30′ each side of the direction.

To make the above described curves definite enough to fix all their parts, the bearing of one of the radials or tangents of each curve should be included, thusly:

    a. "Thence from a tangent line bearing N 38° 13′ E, northeasterly 60.00 feet along a curve concave to the southeast having a radius of 100.00 feet to the beginning of a reverse curve concave to the northwest having a radius of 60.00 feet; thence northeasterly 40.00 feet along said curve."

If the line previous to the curve had already been cited, it would be unnecessary to repeat the bearing because it would be properly assumed that the curve was tangent unless otherwise stated.

    b. "Thence from a tangent line bearing N 29° 40′ E, easterly 60.00 feet along a curve concave to the south having a radius of 60.00 feet to the cusp of a curve concave to the southeast having a radius of 50.00 feet, to which point of cusp a radial of the last mentioned curve bears N 23° 15′ W; thence southwesterly 40.00 feet along said curve."

The last bearing gives orientation to the new curve. The bearings of radials (unless otherwise definitely stated) are expressed in a direction *from* the center of the circle to its circumference. The radial line stops at the circumference of its circle. If you wish to follow a radial line beyond that, you must use the expression, "along the prolongation of said radial line . . . ."

     c. "Beginning at a point on a curve concave to the southeast having a radius of 200.00 feet, to which point a radial line bears N 47° 28′ W; thence northeasterly 60.00 feet along said curve to the cusp of a curve concave to the southwest, to which the center of the circle of said curve bears S 72° 06′ W 70.00 feet; thence northwesterly 40.00 feet along said curve."

The use of the bearing and distance to the center of the circle accomplishes two objectives in one statement: both the radial line and the radius are given.

     d. "Beginning at a point N 2° 13′ E. 200.00 feet from the center of a circle; thence southeasterly along the curve of said circle concave to the south a distance of 60.00 feet to the beginning of a nontangent curve concave to the northwest having a radius of 75.00 feet, to which beginning of curve a radial bears S 23° 14′ E; thence northeasterly 40.00 feet along said curve."

This time the combined radial line and radius were used to start the description. In all of the above cases, the bearing of the radial to the end of the first curve would be calculated.

With the bearings cited to control one point of each curve, it is possible to draw each pair of curves in exact relation to each other.

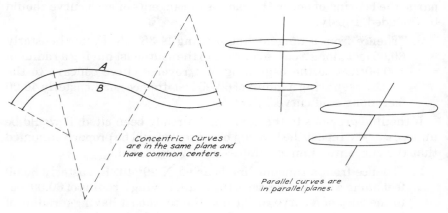

Concentric Curves are in the same plane and have common centers.

Parallel curves are in parallel planes.

Fig. 3-26   Fig. 3-27

Other Terminology

Curves are said to be concentric if they have a common center point and different radii and lie in the same plane. See Figure 3-26.

Curves are parallel if they lie in different planes and their center points are in the same line through the planes. This would hold whether the radii were the same or different. See Figure 3-27.

Two or more curves having the same radius but not common center points (whether in one or more planes) are considered to be equal or equi-distant apart but not concentric nor parallel. See Figure 3-28. Although corresponding points are at equal distances along parallel lines, the curves are still not accepted as parallel.

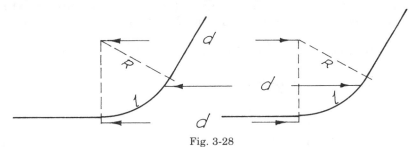

Fig. 3-28

You will note in preceding descriptions of curves the use of the phrase, "along said curve." This expression is necessary in order to connote that the measurement of the line follows the arc of the curve and not across it on the chord.

Contrariwise, if the form were used, "thence easterly to a point on said curve 40.00 feet from the westerly terminus of said curve, "or" easterly 40.00 feet to a point on said curve," this would connote that the distance is measured along the *chord* of that curve and not along the arc.

Multiple Curve Conditions

Sometimes too much information put into a legal description is as detrimental as insufficient information. Great care must be taken in describing curves and their relationship to other lines because it is easily possible to use too many parts of a condition, some of which may not be compatible with certain other ones.

For example, this portion of a description follows curves to a line that is supposed to be both tangent and parallel with, and a certain distance from, an existing record right-of-way:

" . . . . . to the beginning of a curve, tangent, concave westerly and having a radius of 480.00 feet; thence northerly and northwesterly along said curve through a central angle of 39° 38′ 26″ an arc distance of 332.09 feet to a point of reverse curvature (a radial through said point bears N 50° 37′ 34″ E), said point being the beginning of a curve, tangent, concave easterly and having a radius of 480.00 feet; thence northwesterly and northerly along said curve through a central angle of 39° 38′ 26″ an arc distance of 332.09 feet to a line tangent, said line being 41.50 feet easterly, measured at right angles from the easterly right of way line of the Edison Securities Company as described in deed recorded May 23, 1954 in book 2736, page 411, Official Records of said County; thence N 0° 16′ 00″ E along a line parallel with and 41.50 feet easterly of said easterly right of way line . . .″

Note that all three conditions for both curves are cited and then qualified as terminating in a tangent that is also parallel with and 41.50 feet easterly from the right-of-way line. It is highly improbable that all

Possible Alternatives With Multiple Fixed Curve Calls (exaggerated)

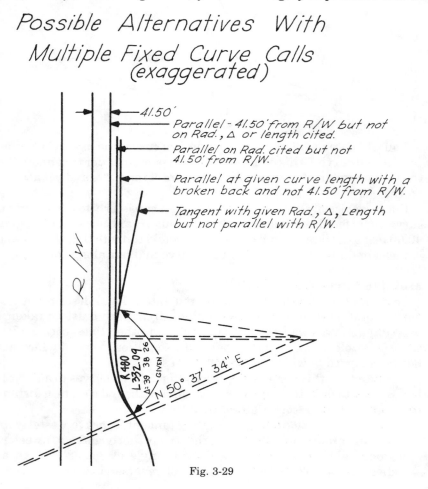

Fig. 3-29

of these conditions will remain intact in the field relocation; it would be remarkable if physical development of the curve concluded in a line that was both tangent to the curve *and* parallel with the right-of-way, *and* 41.50 feet from the right-of-way. See Figure 3-29.

Although you can draw a map that will show these facts, you must realizing that field conditions would most likely render this compatibility impossible. You may do well to find what the survey crew determined in the field and follow their notes in making the map.

Rehabilitation of Curves

In determining which elements should be held in describing or rehabilitating a curve when a field resurvey reports conflicting facts, one needs to know how curves are created. In the field location of the centerline of a road, either an existing one or a new one to be established, the general procedure is to locate the middle of the straight sectors and prolong their tangent lines to intersect and set those points of intersection (PI) for control. See Figure 3-30. The deflection angle at the PI is then measured, although sometimes the angular measurement is set first at an even figure and the tangent line moved to conform. If the ground conditions are unacceptable for setting a point, as perhaps it might be at PI #15 in the illustration, either the radius to the center of the circle, of which the arc will be tangent to the tangents, will be set off and measured or the chord will be established and measured for length and deflection angle. Of course, the deflections from the chord to both tangents must be equal.

After the deflection angle at the PI is measured (which by geometry equals the central angle), usually either the radius is assumed at an even number of feet or the semi-tangent is set at an even number of feet. In either case, the control is established and the remainder elements are calculated from them.

It is because of the fact that two of these three elements (the central angle, semi-tangent or radius) are determined in the field to be the controls for the curve that you look to them first in the event of any discrepancies.

If, for example, the curve data is shown on a map as:

$$R = 220.00'$$
$$\Delta = 35° \ 00' \ 00''$$
$$L = 134.39'$$
$$ST = 69.37''$$

it is obvious that the creator set the angle and assigned the even foot radius.

It the field notes of a resurvey of this curve show:

$$R = 222.31'$$
$$\Delta = 34° 57' 20''$$
$$L = 135.63'$$
$$ST = 70.00'$$

it would appear that the field angle was measured and found to be less, and that an even footage was arbitrarily assigned to the semi-tangent.

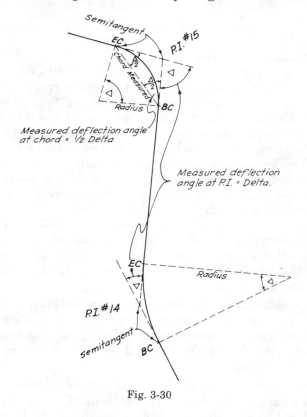

Fig. 3-30

Which statements will hold? Do you accept them as submitted? Going back to the normal method of creation, it would appear that the even foot *radius* should hold and if the remeasured angle is less, as shown, then the semi-tangent and arc length would be calculated from that angle and the held radius, thus:

$$R = 220.00'$$
$$\Delta = 34° 57' 20''$$
$$L = 134.22'$$
$$ST = 69.27'$$

See Figure 3-31

The resurvey should not willfully change the element control as first established. On the other hand, neither should you write your description changing the surveyor's field note information without discussing the matter first. He may claim to have found one or more monuments upon which he based his decision; however, if one adheres to the well-established custom of following in the footsteps of the original surveyor, it would appear contradictory that he would show an even 220.00 foot radius on the map while setting stakes at an incompatible even 70.00 foot tangent.

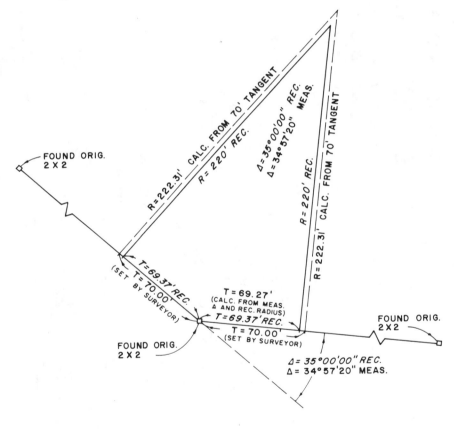

Fig. 3-31

Curve Controls

Another facet of curve control involves the juxtaposition of a curve which is non-tangent to its opposite boundary but tangent to the street line control. In Figure 3-32, such a configuration is shown for an area to be abandoned.

The description can be written in either direction but each way must carry its particular ties in order to hold the correct relationship of the

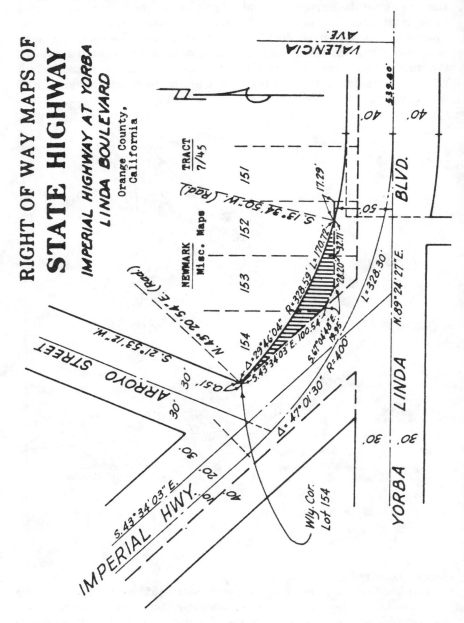

Fig. 3-32

curve to the other lines and you must watch this carefully in order that your map reflects the facts correctly.

Clockwise:

Those portions of Lots 152, 153 and 154 of the NEWMARK TRACT in the county of Orange, state of California as per map recorded in book 7, page 45 of Miscellaneous Maps in the office of the County Recorder of said county described as follows:

> Beginning at the most westerly corner of said Lot 154; thence along the westerly line of said Lot N 21° 53′ 12″ E 0.51 feet to its intersection with a curve concave to the northeast and having a radius of 328.59 feet and which is tangent to a line parallel with and 40 feet north of the center line of YORBA LINDA BLVD. as shown on said map and to which intersection a radial bears S 43° 20′ 54″ W; thence southeasterly 170.72 feet along said curve to a line parallel with and 50 feet north of said center line, a radial to said curve at said intersection bears S 13° 34′ 50″ W; thence along said parallel line S 89° 24′ 27″ W 60.91 feet; thence N 67° 04′ 48″ W 19.95 feet to the southwesterly line of said Lot; thence N 43° 34′ 03″ W 100.54 feet to the point of beginning.

(Note that curve is tangent and 40ft. north of ℄ of Yorba Linda Blvd.)

Counterclockwise:

Those portions of Lots 152, 153 and 154 of the NEWMARK TRACT in the county of Orange, state of California as per map recorded in book 7, page 45 of Miscellaneous Maps in the office of the County Recorder of said county described as follows:

> Beginning at the most westerly corner of said Lot 154; thence along the southwesterly line of said Lot S 43° 34′ 03″ E 100.54 feet; thence S 67° 04′ 48″ E 19.95 feet to a line parallel with and 50 feet north of the center line of YORBA LINDA BOULEVARD as shown on said map; thence along said parallel line N 89° 24′ 27″ E 60.91 feet to a point on a non-tangent curve concave to the northeast and having a radius of 328.59 feet and which is tangent to a line parallel with and 40 feet north of said center line and to which point a radial bears S 13° 34′ 50″ W; thence northwesterly 170.72 feet along said curve to the westerly line of said Lot; thence S 21° 53′ 12″ W 0.51 feet to the point of beginning.

Railroad Curves

A special type of curve to consider is that developed by the railroads based on their field method of establishing the center line for laying tracks. It consists of setting points along the curve at certain *chord* intervals (normally 100 feet) and using a constant deflection angle. This is why such a curve is designated as "a 1° curve," "a 4° 30″ curve," etc.; it means that "1°" is the deflection angle used at every station, etc. The radius is obtained from trigonometrical derivation. The commonly used chord definition is, a 1° curve has a chord of 100 feet subtending a central angle of 1°, while a 1° curve of arc definition (such as is used by some highway departments) has an arc of 100 feet subtending a central angle of 1°. In practice, 100 foot chords are used with deflection angles up to 7 degrees; 50 foot chords are used from 7 to 14 degrees; 25 foot chords are used from 14 to 28 degrees, and 10 foot chords are used for the larger degrees in sharp curves. *The reader is also referred to other standard texts on route surveying for additional information.

An easy way to visualize and remember the relationship between these factors is that a small angle has a long radius and a large angle has a short radius. With this in mind, you can understand how a spiral curve composed of a series of compound single curves with increasing, or decreasing, radii can be used for making an easy transition from a straight track into a curve section and then out to the tangent. Because high speed trains proved that the transition from a tangent line into the old original single simple curve and out to the tangent was too abrupt, the railroads developed and refined special curves in various combinations to be inserted for the transition. These are sometimes referred to as *taper curves* or *transition curves*.

In your use of these curves, remember that practically every railroad company through its engineering department has developed its own formula; therefore, be sure to check on the particular design in use by the railroad with which you are dealing.

Conveyed vs. Described.

Beware when you see the word "conveyed" in a description; it can carry hidden meanings. It may include and "convey" a larger area than described in the words, or, it may actually "convey" less area than described.

The usual form of reference to a boundary deed is, ". . . . to the north line of the land described in deed to P. R. Reynols recorded March 2, 1956 in Book 746, Page 54, of Deeds . . . . " This means that you are tieing to that north line *as it is described* in that deed and properly so. This is the preferable way to use such a reference.

It, on the other hand, you have a reference," . . to the westerly line of the land conveyed . . . ", remember that that may be more or less than described as illustrated in Figure 3-33.

*FIELD MANUAL FOR RAILROAD ENGINEERS, Nagle.

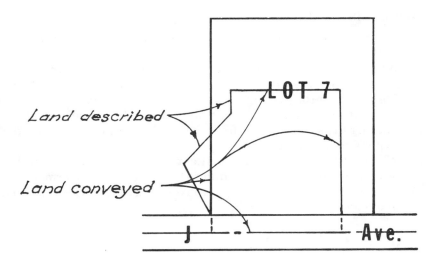

Fig. 3-33

As a note of warning, however, do NOT change the word "conveyed" to "described" if it is that way in an already recorded document; in such a case, you must live with it and search back title to determine its true limits. Otherwise, get a title company's or an attorney's opinion on the extent of the area "conveyed".

PROBLEMS
1. Are all government sections regular?
2. Some sections are fractionized because of a river or lake boundary; are lots within a section called "fractional" lots?
3. Is the meander line of a lot or section adjoining water the true line of ownership?
4. What does a metes and bounds description of a subdivision of a section indicate?
5. What can improve the understanding of a legal description?
6. Are descriptive ties to adjoining land necessary when you are given just a quarter, quarter section description?
7. List 6 items you need to use in the caption on a map.
8. What are the 3 methods used for laying out and developing a map? Describe them.
9. Do you need to mention anything about coordinates when you show them assigned to points on a map?

PROBLEMS, Cont'd.

10. What are the two estates in respect to easements and to which land do they attach?
11. What are the two classifications of easements? What is the difference between them?
12. Describe "overburdening" as it applies to an easement.
13. In creating a strip easement should you be concerned about whether the survey differs from the legal description covering the owner's property? Explain why.
14. Which elements in a horizontal curve will be used to control it when new field information differs from the record information?
15. What is the difference between "conveyed" and "described" as used in a legal description?

Chapter 4

Lettering and Lines

Lettering and Lines

   Legibility is the keynote in lettering. From the fundamentals of the strokes involved in forming letters, there have been thousands of variations made at the whims of individualists, most of which retain legibility but some of which veer away enough to make some of the forms questionable.

   Aside from knowing the ways of forming letters, there is only *one* way to achieve the ability to make good lettering — that is by practice. Not until the muscles in the arms, hands and fingers become accustomed to the pattern of movement necessary to make the correct strokes that formulate the letters do those muscles get "in the groove" and produce beautiful effects.

Besides short lines and curves used for lettering, there are other forms of lines used for representing different things, such as pipelines, street lines, contours, etc., and they are used in various widths, weights and combinations of long and short dashes to give distinctive representation. These line expressions are accomplished through the controlled use of pencils and pens.

Fundamentals in Letters

One studying the history of letter-graphics would find a very interesting array of forms. Over the centuries, and even back in time to civilizations heretofore not reported in history (though sometimes in mythology), the hieroglyphics uncovered by archeologists reveal endless forms of scribing to represent thoughts in line.

Today the fundamental combinations and forms of the lines and curves used for the twenty-six letters and ten basic figures in the English language have become stabilized but within that framework, there are hundreds of variations. This can be seen in a review of any typographer's offer of fonts of type; however, there are four general classifications: 1, Roman; 2, sans serif; 3, script; 4, black letter.

Roman is most familiar for its elegance expressed with serifs, or finials, and contrasting thick and thin strokes. Roman in its original form was vertical and typographers today classify vertical type as roman; however, Funk & Wagnalls *Standard* Dictionary states, "Both Old Style and Modern types are made in the vertical form (roman) and most of them are provided with an accompanying form (italic), a sloping letter introduced by Aldus Manutius, Italy, 1501, and used principally for emphasis." Typical representations are found in the types known as Century (used in this book), Baskerville, Bodoni, Caslon and Garamond.

Italic (pertaining to parts of ancient Italy, other than Rome) refers to type that slants up toward the right. It was developed from the writings of Petrarch, an Italian poet and first printed in an edition of Vergil in 1501 which was dedicated to the states of Italy. It is faster to execute than the vertical letter.

D. M. Anderson in his book, "Art of Written Forms", classifies Roman as follows:

Old Style Roman: capitals, lowercase.
Old Style Roman italic: capitals, lowercase.
Modern Roman: capitals, lowercase.
Modern Roman italic: capitals, lowercase.

One form of Roman vertical and Roman italic are shown in Figure 4-1.

abcdefghijklmnopqrs 1234567890
ABCDEFGHIJKLMNOPQRSTUV
Roman vertical

abcdefghijklmnopqrstuvwxyz 1234567890
ABCDEFGHIJKLMNOPQRSTUVWXYZ
Roman italic

Fig. 4-1

The sans serif type is primarily a uniform width stroke letter without the flare, although there have been a few fonts made with slight differences of widths leaning toward the thick and thin style without the serif but with a very slight flare at the end of each stroke. D. M. Anderson further states, "Sans serif alphabets are designed in an upright version and also in a version slanted to the right. These slanted versions are incorrectly labeled italic . . . . . italic stands for a version of Humanistic minuscule developed out of pen letters of the fifteenth century in Italy." He says the term *oblique* is preferable for slanted versions of sans serif alphabets. Sometimes the sans serif is considered as a modern gothic. An example of the vertical and oblique is shown in Figure 4-2.

abcdefghijklmnopqr 1234567890
ABCDEFGHIJKLMNOPQRSTU
Sans Serif vertical

abcdefghijklmnopq 1234567890
ABCDEFGHIJKLMNOPQRSTU
Sans Serif oblique

Fig. 4-2

Script is the nearest form to writing and one style is illustrated in Figure 4-3.

abcdefghijklmnopqrstuvwxyz 1234567890
ABCDEF GHIJKLMNOPQRS
Script lettering

Fig. 4-3

The black letter, a so-called compressed style, resembles German manuscript handwriting and is exemplified in the Fraktur, Old Gothic, Old English, Text and Scwabacher types.

With each form of type, there is usually a variation in the weight of the lines used in the letters: heavy line letters are called "bold" or "bold face"; light line letters are called "light" or "light face"; and in between these are the medium weight line letters called "medium face".

The thick-and-thin aspect of the roman letters was developed from the use of chisel shaped quill pens. The chisel point was generally held in one position which resulted in all strokes made in one direction were thick and those made in the other direction were thin. For example, follow the creation of the letter V in Figure 4-4: the stroke on the left side drawn from top to bottom was made with the chisel edge broadside while the right-hand stroke was made with the thin edge. The serif is added afterward making the finished letter. From this procedure the draftsman can determine which stroke of any letter should be thick and which one thin.

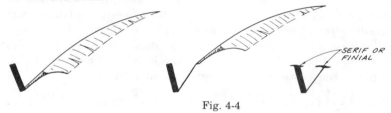

Fig. 4-4

Roman lettering, either vertical or italic bold face, is reserved for use as an impressive conveyance of ideas such as in advertising, motion picture titles and posters, titles of books and headings of chapters, etc. Under the subject matter of this book, it would apply to names or numbers of subdivision maps, titles of right-of-ways or projects, street names, etc.

Sans Serif lettering, either vertical or slant, is used a great deal because of its simplicity, esibility of execution and readableness. For these reasons, it is the most common type used in keeping field notes.

Script lettering takes more time to make it look regular and therefore it is not as commonly used as the others.

Let us now consider the basics of forming letters. Sometimes the things we see every day are taken so much for granted that we miss the interesting details. You realize in a casual way that some letters are high and some are low, while others extend below. To designate and control the relationship of these vertical differences, the guidelines are labeled:

$$Lg \begin{array}{l} \text{------------ cap line} \\ \text{_____waist line} \\ \text{----------} \\ \text{_____base line} \\ \text{----------- drop line} \end{array}$$

These are set at different spacings to achieve different effects. Although the proportions of the vertical height of the waist line range from three-fifths to three-fourths of the distance from base line to cap line, the circular type lettering guide is punched for either two-thirds or three-fifths. The drop line is the bottom limit for such letters as g and y.

Besides the Ames Lettering Guide shown in Figure 5-8, there are also punched-hole guides available in triangles, one of which is shown in Figure 4-5.

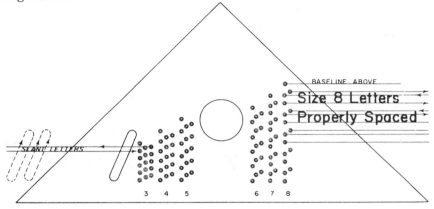

Use of lettering triangle
Fig. 4-5

Note that in this example, no drop line is illustrated. In the circular type guide, a drop line can be obtained by using the first, second, fourth and fifth holes in the center column.

If you are working on transparent material, you can place an equilined sheet under it for a lettering guide. The other way is as shown here; lay a ruler at whatever angle will give you the spacing interval required, tick-mark the units and draw the horizontal lines in light pencil.

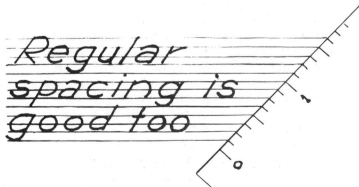

Slant guidelines are usually set at either 68° or 75° from the horizontal.

A good way to help your muscles get into the correct pattern for the various strokes of letters and figures is to spend at least ten minutes each day, and more if possible, doing the following exercises. You may use cross-sectioned paper or lined paper. For the practice on slope lines, superimpose lightly drawn slant lines about one-half inch apart.

First, do a complete series of vertical lines across the page with down strokes from one line to the next one below it, thus: ▓▓▓▓▓▓▓▓

Next, do a series of slant lines across the page, thus: ▰▰▰▰▰
and thus: ▰▰▰▰▰

Then, make blocks of horizontal lines across the page, thus:
≣ ≣ ≣ ≣

To get the feel of the circular or elliptical pattern, draw the "spring" exercise across the page, first clockwise and then counter-clockwise, vertical and slant, thus: ⟋⟋⟋⟋⟋ and thus: ⟋⟋⟋⟋⟋

You will find that after a certain amount of repeating these exercises, you will be able to draw the fundamental lines and curves without irregularities. In time, the straight lines, proper slants and curves will become natural and automatic. The more one practices this preparation for lettering, the sooner will he manifest a professional appearance in his lettering.

As a matter of nomenclature, capital letters are also referred to as "upper case" letters and small letters are also known as "lower case" letters. These designations were assigned because in the printer's shop, the capital letters of each font of type were kept in an "upper" drawer of the case and the small letters were kept in the "lower" drawer immediately below the "upper" one.

For the construction of letters, we will begin with each form of the simple vertical single line sans serif type capitals as shown in Figure 4-6 which also shows the direction to draw each stroke.

Because certain letters appear "blacker" than others because of either the number of strokes or the closeness of the strokes, they in particular need to cover a little more space than others. Also in order to maintain even "color" along a line of lettering, the letters must have certain spacing between them to balance the blacker against the lighter ones. Consider the following phrase with close spacing:

MEN NEED TO LIVE OUTDOORS.

If you look at this line with your eyes squinted, you will notice how part of it has a much darker color than the rest, whereas with proper spacing applied between the heavy letters, it presents a more even color, thus:

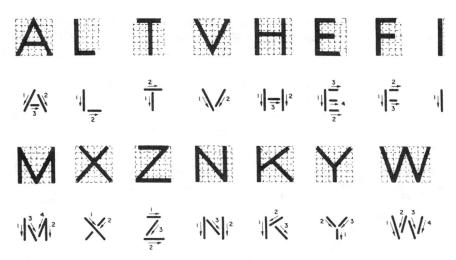

Vertical straight-line capitals.
Fig. 4-6

MEN NEED TO LIVE OUTDOORS.

To use the identical amount of spacing between all letters creates an unbalanced appearance as evidenced in **illiteracy or LAUGHING** whereas proper spacing gives a more pleasing and easier to read effect, thusly: **illiteracy and LAUGHING.**

Similarly, there should be spacing between words so they do not run together and become difficult to read. The space of an "o" between words makes for easy reading and when there is a comma, use the same space between it and the next letter.

In order to maintain a balance of color which makes easier reading, not only is spacing used judiciously but the extent and direction of certain strokes of letters are adjusted to accomodate the requirement. For instance, the horizontal stroke on the T or the L may be lengthened or shortened to balance space with the adjoining letter. If T precedes or follows A, the top stroke may be extended slightly but if it precedes or follows I, the top stroke should be shortened so as not to leave too much space between the two vertical strokes. Likewise, if L precedes I or A, the lower stroke should be shortened slightly but if it precedes V or W, the bottom can be extended a little.

Another way to even the color is to change the width of such letters as A, V, W, M, X, N, K, R, B, S, Z, etc. One in particular is the M; by changing from the block type M to the upside-down W type $\wedge\!\wedge$, the bottom spacing between the legs renders a more even spread of color.

It is not desirable to make the top stroke of the letters T, E, F and Z too long because that gives the feeling of top-heaviness and unbalance.

When pairs of letters have their vertical strokes next to each other, such as *I* and *L*, *H* and *E*, *N* and *K*, etc., they should have a little more space between them than you would have between such combinations as *I* and *O*, *D* and *O*, *I* and *V*, *A* and *L*, etc.

In the case of pairs of slope stroke letters, such as *A* and *V*, do not place them too close together nor too far apart either.

Let us now consider those vertical sans serif type capital letters with curve strokes as illustrated in Figure 4-7 also showing the direction to draw each stroke.

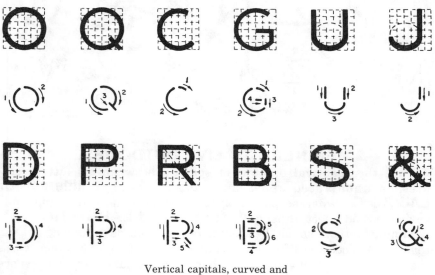

Vertical capitals, curved and
straight-line combinations.
Fig. 4-7

The most difficult maneuver is to make freehand circles or ellipses symmetrical. Although the following procedure may take a little more time and effort in the beginning, you will be rewarded with more perfectly formed and better looking letters. The two-stroke method of making the *O*, *Q*, *C* and *G* letters as indicated above is the ultimate goal but for simplicity in acquiring symmetry, start with four strokes as shown in Figure 4-8. You may even wish to draw the ─┼─ first and mark ticks on each line equidistant from the center. After some practice with this, you will find you can produce a satisfactory form by combining strokes 1-and-3 and 2-and-4.

A similar procedure for the letter *S* can also simplify the accomplishment of making good forms. The *S* is in essence a part of the figure *8* which in turn is composed of either two circles, or two ellipses, one atop the other, the top one being slightly smaller than the bottom one. In the beginning use the four-stroke method outlined above and draw the lower circle with a diameter or height equal to three-fifths of the total height, then draw the upper circle in the remaining height by overlapping the parts of the two circles which are common. Also practice this using the four strokes in *elliptic* form with the long axis horizontal and the lower ellipse between one-fourth and one-third longer than the upper one. See Figure 4-9.

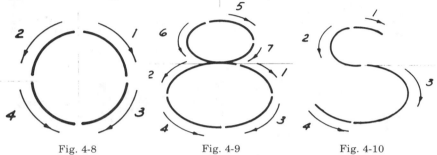

Fig. 4-8   Fig. 4-9   Fig. 4-10

After practicing the formation of the figure *8*, it is recommended you try the letter *S* in the four steps illustrated in Figure 4-10. This will resolve itself later into the three strokes shown in Figure 4-7.

Although the style shown above for the letters *G, P, R* and *B* is typical block form, a better sense of balance will be seen by making the following changes. If the *G* circle is stopped at the bottom and a tangent line is drawn to the right to a point just below the beginning of the top with the vertical leg projected down to it, thus: G, the letter is given more stability.

If the lower arm of the *P* is dropped one-half square and the semicircle is moved one square toward the stem, thus: P, it will not have such a top-heavy feeling.

It has not been shown here but sometimes the letter *A* is formed with a square top, thus: A, and because of this, the letter *R* with the slant leg joining near the upper circle, especially when drawn hurriedly, takes on the appearance of the square A, *thus* R. Consequently, for safety, it is better to use the type of *R* that attaches the slant leg at the intersection of the stem with the bottom of the upper part. Along with this goes the same recommendation for a modified top similar to the one described for *P* making it look thusly: R.

In the same way that the *S* and *8* look better with the top portion smaller than the bottom, so does the *B* look better balanced with a smaller top, thus: B. This cannot be mistaken for a square A either.

Next, look at the construction of the lower case letters shown in Figure 4-11. Note the addition of the "DROP LINE" to take care of the bottoms of *g, j, p, q* and *y*.

The crosses of the *f* and *t* are made at the "WAIST LINE" and note that the top of the *t* is short and does not go to the "CAP LINE."

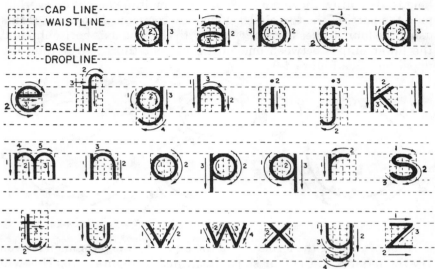

Vertical lowercase letters.
Fig. 4-11

All of the letters with circles and lines combined have the tangent line *coincident* with the circle, not outside of it.

Again remember that the position of the "WAIST LINE" is not rigid (p. 4.4) but having once decided where you want it, keep it there throughout that particular job.

In the construction of numerals, Figure 4-12, you see the use of ellipses in lieu of circles. This is where your "spring" practice (p. 4.6) plus the strokes for a figure *8* and letter *S* (p. 4.9) will give the form to use in drawing them.

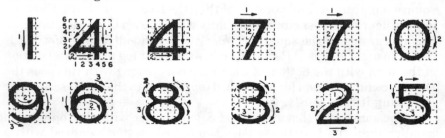

Vertical numerals.
Fig. 4-12

It is mandatory that numerals be distinct because confusion can result in misinformation and sometimes costly problems. Perhaps the two pairs of figures most commonly mixed are the 7 with the *1* and the 5 with the 6. The reason for the 7 problem is that the top stroke is made too short or not horizontal, thus: *7*, therefore, it looks like a *1* with a flag. Obversely, a *1* with too long a flag on it, thus: *1*, can look like a short 7. The 5 problem results usually from insufficient straight lines and too long a curved tail, thus: *5*, which at a glance has the similarity of a 6. One way to draw a figure 5 so as to keep it from resembling a 6 is to drop the tail, thus *5*. In addition, the form of 3 using a miniature 7 on the top, thus: *3*, sometimes done hurriedly results in either a *3* or a *5*, and thereby becomes mistaken for a 5 so keep the 3 well rounded.

There are two ways of drawing fractions, either by placing the numerator and denominator above and below a horizontal line, thus: $\frac{1}{2}$ or by placing the numerator first followed by a slash line and then the denominator, thus: 1/2. The first way occupies a vertical space equal to twice the height of a whole number while the second form may equal the height of a single number or be slightly reduced. In between these two methods is a compromise form still using the slash line but which elevates the numerator and depresses the denominator enough to make their total vertical height equal to about one and a half times the regular figure, thus: ½.

The next consideration is the inclined or oblique letters and figures. The same strokes and comments apply as previously outlined, but with the slanted orientation as indicated in Figure 4-13. All of the circles become ellipses and their axis must be at the *same* slope throughout the job whether it be 45° or some other angle.

Inclined letter formation.
Fig. 4-13

With certain letters, regardless of type, size or slant, extensions are needed in order to create the optical illusion of conformity. For instance, when pointed capital letters like *A* and *V* are set next to square type letters such as in the word DAVID , they seem to be short so in order to give them the illusion of proper height, the points are slightly extended thus: DAVID . This applies also to the points of *M, N* and *W*.

The same principle applies to rounded letters. For example, in the word TOOK , the *O* seems lower or shorter than the *T* and *K*; therefore, in order to give them the illusion of proper height, the circle or ellipse as the case may be is extended slightly above and below the line, thus: TOOK . This applies also to C, Q, S, 3, 6, 8 and 9, to the upper part of 2 and G and the lower part of U and 5.

The matter of spacing slant letters requires the same consideration and application as with the vertical.

Having followed the basic principles of design, execution and spacing for sans serif letters, you can now expand into their application to the roman type lettering by using the thick and thin strokes with serifs. Unless you use the chisel form of pen or brush, you will need to make double strokes for the wide lines and either leave them open, thus: A , or fill them in solid, thus: A .

Types of Letters.

As stated before, there are thousands of different styles of lettering and it would be impractical to illustrate or review them all here; however, the following will give you some idea of a few of the variations.

In the development of so-called modern sans serif, all of the horizontal lines and points of intersection of angular lines are placed either high, thusly: *HAEFGKRB* or low, thusly: *HAEFGKRB* .

A combination of both is also used, thusly: *ABEFGHKR* .

Cursive script letter is an approximation to handwriting because it gives the effect of joined-together letters, but in its true form it uses thick-and-thin strokes rather than the single thickness so it is not commonly used in survey drafting because of the extra time required to follow the form.

The element of time is equally a deterrent in the use of roman serif type lettering but you can see from the following how impressive it can be for title blocks, report headings, etc.:

Special Report on Curves

Drafting Demands
for
Astronaut Preparation

HIGHWAY JUNCTIONS!

Even gothic in a bold face can be impressive:

EDM Surveys in Alaska

The conclusion to be drawn here is that for most of the survey drafting work, and this applies to field notes too, the sans serif oblique, otherwise known as single stroke slant letter, is the easier, faster, more legible lettering to use while the roman type is reserved for special applications.

Lines.

Whereas we have been discussing the use of lines for the creation of letters, there are other needs for lines to which a combination of weights and characteristics have been assigned by which to express through symbols certain established definitions.

Weights

As to weight of line, four different widths are recommended: *thin* (1/100 in or 0.35 mm) for center lines, dimension lines, leaders, long

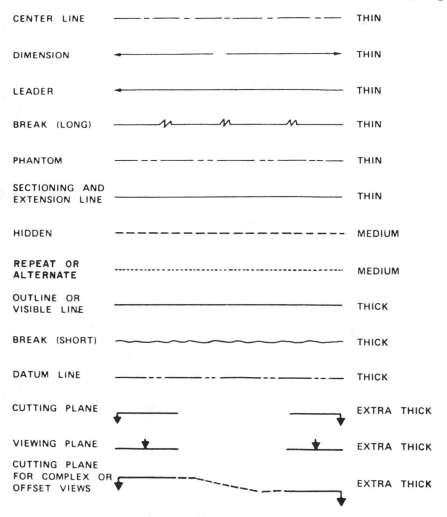

Fig. 4-14

breaks, sectioning extensions, phantom and reference lines; *medium* (1/60 in. or 0.423 mm) for hidden lines, repeat and alternate lines; *thick* (1/40 in. or 0.635 mm) for outline or visible lines, short breaks, datum lines; *extra thick* (1/25 in. or 1.016 mm) for cutting plane and viewing plane lines. See Figure 4-14. These widths are not rigid but to give you relative values.

While it is possible to adhere to the above thicknesses fairly well with either the ruling pen or the new technical fixed-diameter pens, lines drawn with pencils will be generally thinner. An alternative with pencil drawing would be to use different hardnesses in lieu of different widths such as: 4H for *thin;* 2H for *medium;* H for *thick* and B for *extra thick* or any comparable grouping: again it is a matter of relativity.

Characteristics

Now that the *weights* of lines have been given some definition, let us discuss the *characteristics* assigned to forms of lines which convey certain meanings. Figures 4-15, 4-16, 4-17 and 4-18 show the applications of these lines.

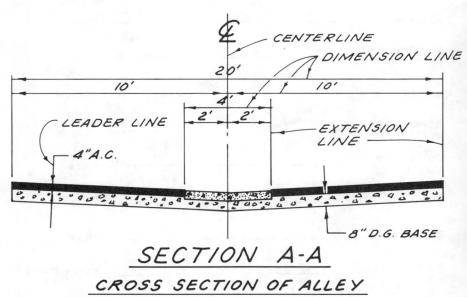

Fig. 4-15

Centerlines are composed of long and short dashes alternately and evenly spaced with the long dash generally at each end; at intersections, the short dashes intersect. This pattern may be broken for centerlines of very short length if it does not result in confusion.

Dimension lines, as the name implies, are used to indicate the extent of a measurement between the points of the arrowheads on each end. Usually the dimension line is unbroken with the distance shown above the line on construction drawings, such as plans and cross sections, but the line is broken with the measurement placed within the break for production drawings of machinery, etc.

A *Leader line* directs the eye from pertinent information to the part or portion to which the note, number or reference applies and the line ends in either an arrowhead or dot. Arrowheads most always terminate at the line while dots are within the outline of the object. Penetration of the leaders is permissible when necessary for clarity.

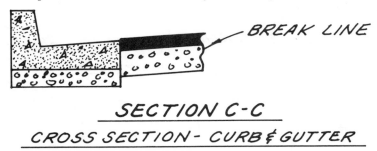

SECTION C-C

CROSS SECTION - CURB & GUTTER

Fig. 4-16

Break Lines are used to indicate a break in the drawing of an entire object by means of shortening the drawing or by cutting off a part of the drawing. It never changes the actual length given in the dimension. The long break line made with freehand zigzags ⌐———⌐ indicates a part left out. The short break drawn freehand ⌡ or ∿∿∿ indicates the other part left off as in Figure 4-16.

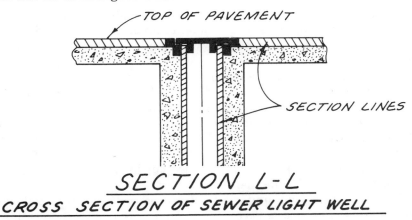

SECTION L-L

CROSS SECTION OF SEWER LIGHT WELL

Fig. 4-17

Phantom lines are used to show an alternate position of any part, the relative position of an absent part or a repeated detail. They are composed of alternate one long and two short dashes evenly spaced with the long dash at each end.

Sectioning lines are usually drawn diagonally to indicate exposed surfaces in a cut-away or cross-section view as in Figure 4-17.

Extension lines are projected out from the limits of whatever item is to be dimensional and they are the lines to which the arrowheads on the dimension lines point. Extension lines are not to touch the subject outline. See Figure 4-15.

Hidden lines are short dashes evenly spaced to show features that are hidden in the delineated view. See Figure 4-18.

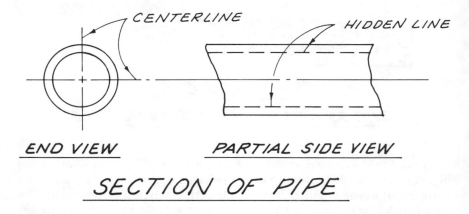

Fig. 4-18

Repeat or *alternate lines* composed of very short dashes are used similarly to phantom lines but in smaller areas or in a limited space.

Outlines or *visible lines,* as the words imply, are used for the actual outline of the subject at hand.

Datum lines indicate the position of a base plane of elevation from which other elevations are derived. The datum line is composed of one long dash and two short dashes evenly spaced.

Cutting-plane and *viewing-plane lines.* Cutting-plane lines indicate the plane in which the section is taken. Viewing-plane lines indicate the plane from which a surface is viewed. Where it is necessary to show the view or cross-section as bent or offset, the cutting plane is likewise bent or offset in a parallel manner.

On any drafting job, uniformity of expression along with following conventional line use is necessary so that only one interpretation is possible.

In any situation where there is a coincidence of lines, the following precedence of lines should be followed:

a - Object line.
b - Hidden line.
c - Centerline or cutting-plane line.
d - Break line.
e - Dimension or extension lines.
f - Sectioning or crosshatch lines.

In other words, if a dimension line coincides with a break line, the break line has priority and should be drawn leaving the dimension line out.

Arrows

Although the NORTH arrow for a map can be placed anywhere on the sheet, it should have the north *point* to the top of the page, to the left of the page, or, if the drawing will fit better on the diagonal, then the arrow should point in some part of the arc between the top of the sheet and the left side of the sheet. In rare cases, the arrow may be required to tip to the right but do not allow it to do so to any great extent of angle.

To maintain conformity, the lettering should read from the top of the page or from the left side; that is, if you rotate the sheet's left side 90° to the top, then that part of the lettering would be right side up.

As to the matter of numbering sections, if the job at hand involves only a few as shown in Figure 4-19, there may be a question as to the direction of facing the numbers.

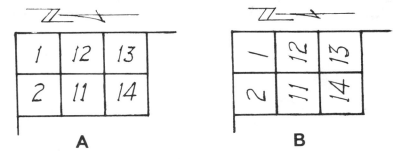

Fig. 4-19

With the NORTH arrow to the left, placing the numbers in the Sections as illustrated in the "A" diagram would conform with principle but the method shown in the "B" diagram is more familiar by association in memory of the way they appear on Official Plats; therefore, the "B" method could be preferable.

As for lettering which is upside-down to the NORTH arrow, it is unacceptable except in a *very unusual* situation.

Legibility has been stressed, and rightfully so, but if comprehensibility is strained or disrupted, the *value* of your drawing can be questioned.

In the case of multiple dimensional arrows, it is advisable to set the arrows back from the point so that they do not overlap or crowd each other. Furthermore, if the lines of the arrows are so many as to be crowded at the point, a small circular space around the point will help not only in clarification at scale but also in the case of reduced size reproduction. See Figure 4-20.

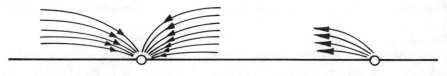

Fig. 4-20

PROBLEMS

1. What is the keynote of lettering?
2. What are the 4 principle types of lettering?
3. What is the distinguishing characteristic of one particular type over the other?
4. Give the names of the 4 lines that control the forms of small letters.
5. Why should one be concerned about the weight (thickness) of lines on a map? Explain.
6. Give 6 examples of line characteristics and explain their use.

Chapter 5

Drafting Tools

Along with knowledge of the methods and application of the principles of the job to be done, it is a well-known fact that the equipment with which one works determines the refinement and sufficiency of the end product. Also the selection and use of the most suitable tools for the execution of your drawings will determine the ease and speed by which you will complete them.

Whereas the man who makes the pictorial story in the field must produce a legible report that combines pictures and words to tell the complete set of facts using a minimum set of tools to produce satisfactory results within certain limits of time that will permit him to keep up with the field crew, the one in the office has not only more time but the accessibility to more and better tools and equipment. The following look at drafting tools will give the reader some idea of a desirable selection with which to work.*

Pencils, scales, triangles, straight edges, pencil pointers, erasers, eraser shield, dust brush and lettering guide constitute the first group of things to use.

Pencils

Whether you use wood-cased drawing pencils or metal holders to receive the leads of various hardnesses, as shown in Figure 5-1, is a matter of personal choice. Much more storage room is required to keep a stock of wood drawing pencils than the small boxes or tubes of leads. Furthermore, in sharpening the wood pencils, you need not only a pencil sharpener (or knife if you are good at whittling) but also a sandpaper pad or steel file for shaping the lead point, whereas with the lead holder, you need only a lead pointer such as shown in Figure 5-2. Furthermore, you must be very careful not to spill the black powder from shaping the pencil onto your drawing.

Lines are drawn in different weights, or color-of-line if you will, to illustrate various conditions; consequently, pencils of different

Lead Holder
Fig. 5-1

Tru-Point "Quick Change"
Variable Taper Lead Pointer
Fig. 5-2

*Photos by courtesy of Keuffel & Esser Company.

hardnesses are used. Soft leads are indicated by the letter "B" with increasing numbers (2B, 3B, 4B, etc.) representing increasing softness and therefore blacker lines, while the hard leads are known by the letter "H" with increasing numbers (2H, 3H, 4H, etc.) representing increasing hardness and therefore lighter lines. The texture of an "F" pencil is in between a "B" and an "H."

Scales

Drawings are made to a certain scale, which represents the ratio between the actual thing and the drawing of it. Usually the drawing is a reduced scale but in some cases it is enlarged beyond actual size in order to show details. Scales are classified as either metric or english and the latter is divided into architectural, engineering and mechanical groups. Here we will discuss the architect's, engineer's and the metric scales.

The most commonly used scale in survey mapping is the engineer's twelve-inch long triangular scale which is divided decimally into ten units per inch, 20 units per inch, 30 units per inch, 40 units per inch, 50 units per inch and 60 units per inch; this is usually designated as a "10, 20, 30, 40, 50, 60 scale." See Figure 5-3.

The triangular type is available in either a flat or concave form, the latter being the lighter weight (see page 2.10). There are also flat beveled scales available in the above units in pairs such as a 10-50 or 20-40 or 30-60 combination.

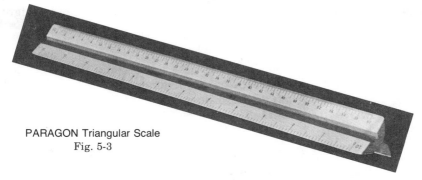

PARAGON Triangular Scale
Fig. 5-3

Other unit divisions, such as 80 per inch, 66 per inch or even 2 inches per mile, etc. are available for special projects. The advantage of the decimal type division is that a 40 scale, for example, can be used so that one inch represents either 4 feet, 40 feet, 400 feet or 4,000 feet.

Drafting machines use the two-bevel flat scales with chuck plates for attachment to the machine head as shown in Figure 5-4.

Metric scales are commonly 300 mm in length and are usually obtainable in graduations as follows: 1:1, 1:2, 1:5, 1:10, 1:20, 1:33-1/3, 1:40, 1:50, 1:80, 1:100, 1:200,

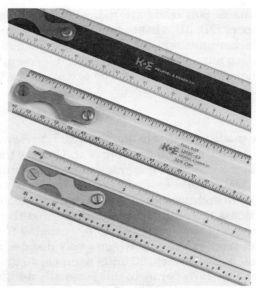

PARAGON Drafting Machine Scales
Fig. 5-4

Triangles

The most commonly used triangles are the 30°-60° and the 45° angles opposite a 90° corner, Figure 5-5. These are available in sizes from 4 inches to 18 inches. The length is measured on the long leg of the right angle, not along the diagonal.

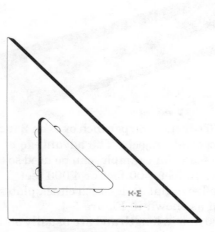

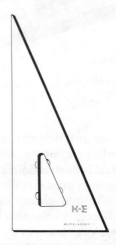

LUXYLITE - 45° Triangle   LUXYLITE - 30° - 60° Triangle

Fig. 5-5

The adjustable triangle shown in Figure 5-6 is helpful in many instances because with the partial protractor, drawing a long line directly from the angular setting is easy without shuffling tools.

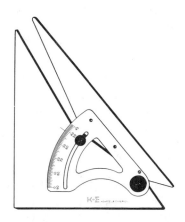

Adjustable Triangle - 10 in.
Fig. 5-6

Straightedges
The execution of long straight lines is done by using what is commonly known as a "straightedge." It can be either a T-Square which is held tightly against the side of the drawing board or drafting table, a Parallel Straightedge (sometimes called a Parallel Rule) which operates in a position always parallel with a predetermined line with cords fastened to the board or table, or a Drafting Machine attached to the board or table, which not only has a protractor by which the ruler attachments can be set at different angles but by which either parallel lines or extensions of lines can be drawn from the desired setting.

Erasers
Erasers are a necessary adjunct to drafting and they come in many shapes and sizes to accommodate not only the hardnesses of pencils but also the easibility of manual application by the draftsman. The electric power eraser is principally a timesaver but must be used carefully so as not to spoil the drawing material.

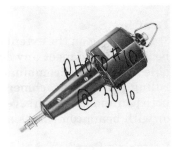

Eraser - TAD   Erasing Shield
Dust Brush

Heavy Duty Electric Eraser

Fig. 5-7

The eraser shield protects the area surrounding the erasure from being smeared or eradicated. See Figure 5-7.

And of course you need a dust brush to clean off the erasures.

Lettering Guides

The expert draftsman with years of experience can do beautiful lettering freehandedly but others need guide lines to maintain uniform heights and uniform slopes. In the old days, guide lines were meticulously (and you could also say tiresomely) drawn at the particular interval format required for the job at hand. Subsequently, holes for various patterns of line separation were punched into a 30° - 60° or 45° triangle and with a sharp pencil inserted into the hole and sliding the triangle along a straightedge, lines would be drawn at the desired intervals.

The Ames Lettering Guide Shown in Figure 5-8 gives a wide choice of line format through three lines of interval pattern-holes on a turnable disc.

Where you are drawing on transparent material, printed guideline forms can be placed underneath, thus saving the trouble of marking your sheet with lines.

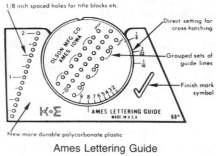

Ames Lettering Guide
Fig. 5-8

The next group of things to consider is drafting tools, pens and lettering sets.

Drafting Sets

Depending on the extent of your activities, your drawing tools should be sufficient to meet any demands of your job. Drafting sets are available in all sorts of combinations from a mere bow compass to a compass, ruling pen and attachments and up to compasses in various sizes for both pen and lead with several different size ruling pens (See Figure 5-9) or with beam compasses as shown in Figure 5-10.

Lettering Sets

In contrast with old metal pen points with various flexibilities for the different weights and styles of lettering, the new type of fixed line width pen with its own ink well makes it possible to have both lettering and lining of uniform character for the many aspects of a project. One type of such pen sets is shown in Figure 5-11 with widths from a 0.13 mm to 2.00 mm.

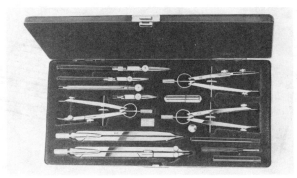

MARK I Three Bow Drawing Set in Corium Case.
Fig. 5-9

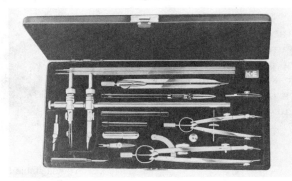

MARK I Drawing Set with Beam Compass in flap-type case.
Fig. 5-10

LEROY 9 Pen Dry Seal Selector Set, stainless tips.
Fig. 5-11

To carry the idea of uniformity into lettering itself, guides with scribers holding a pen to follow the particular style of lettering template desired are also available in sets, one of the most common of which is the LEROY, shown in Figure 5-12.

LEROY Lettering Set.
Fig. 5-12

The adjustable scriber itself is shown in Figure 5-13.

To keep the pen points clean so that the ink will flow smoothly and evenly, a pen cleaning kit is available and shown in Figure 5-14.

LEROY Adjustable Scriber
Fig. 5-13

LEROY Cleaning Kit
Fig. 5-14

Other sets of reservoir pens are available with as many as sixteen points sized from .008 inch to .250 inch.

The next group of tools includes dividers, protractors, curves and planimeters.

Dividers

In a drafting box there may be a separate set of dividers or the compass may be used as a divider by inserting the metal pointer in place of the pen or lead. These are simple dividers and are used for duplicating or transferring dimensions, or dividing straight or curved lines into equal segments.

A proportional divider shown in Figure 5-15 is, as the name implies, proportional one end to the other. They are used for enlargement or reduction by setting the center control along the slot to obtain whatever ratio is desired. For instance, if one set of measurements is to be transferred in a multiple of four times, then the control unit in the slot is set at 4. This type of divider might be called a computational instrument.

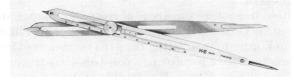

PARAGON Universal Proportional Dividers
Fig. 5-15

Protractors

These are circles or semicircles divided into lines marking 360° or 180° respectively and usually with ½° lines in between. Occasionally, as shown on the Adjustable Triangle in Figure 5-6, only a small part of the 360° circle is marked and used for the particular segment required.

In the case of drafting machines, a vernier scale is attached outside of and adjoining the edge of the protractor circle which permits setting or reading to five minutes of a degree with possible estimating to 2½ minutes.

Curves

Programmed combinations of portions of mathematical curves — such as spirals, parabolas, hyperbolas, ellipses, etc. — are die-cut to produce what are commonly known as "French Curves." Five of the hundreds of combinations are shown in Figure 5-16. These are considered as irregular confined curves.

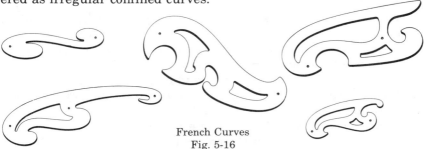

French Curves
Fig. 5-16

The extended curve is shown in Figure 5-17 and can be adjusted to fit
any number and any combination of points within its
length (12″ to 30″) and down
to a three-inch radius.

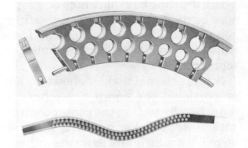

Flexible Curve Rules
Fig. 5-17

Polar Planimeters

These are used to acquire the areas of irregular shapes and they are
available in various sizes to accommodate the project at hand. A small
one is shown in Figure 5-18. Getting the area by means of following the
irregular perimeter with a pointer, or a point-in-glass tracer, is based on
the application of adding consecutive minute slivers of areas of different
lengths. The precision is dependent upon the size of the instrument, the
scale of the map and the skill of the operator.

Polar Planimeter
Fig. 5-18

PROBLEMS

1. Explain the scales of softness and hardness as applied to drafting
   pencils.
2. Name the 3 classes of scales used in survey drafting and cite the
   units available on the most commonly used one.
3. Explain the adjustable triangle and its advantages.
4. Name the 3 kinds of "Straightedges" and tell how they are used.
5. Describe 2 types of lettering guides.
6. Explain the principle and use of a proportionate divider.
7. Explain the principle of the polar planimeter.

Chapter 6

Production of Topographic Maps

Topographic Maps

There are four kinds of topographic information, namely: culture, vegetation, relief and hydrography. The last two have to do with related elevations within limited areas of which the first is above sea level while the other is below sea level.

Again, the boundary of the property under consideration must be correctly established before doing the work on the inside area. Following this, the control stations and traverses from which the topographical features are measured must be oriented to the boundary. If an open traverse is used and no specific tie is made to a perimeter point, then some object on or near the boundary which is sighted and recorded from the last station should be plotted to verify your positioning of the traverse. If it does not fit, find the mistake and rectify it before proceeding.

If a closed traverse is used, it should be checked first for mathematical closure and then for its correct juxtaposition with the boundary.

After these controls are proven satisfactory, the information obtained from each station reading is plotted on the map and contour intervals are interpolated between the marked elevations.

Orthophoto composite maps are being used in many ways because of the variety of combinations possible which makes them adaptable to various requirements. The orthophoto developed from photogrammetry is basic; from this, the contour map is created through the plotter and topography may be included or put on a separate sheet. While one transparency may be made of the property lines, streets, alleys etc. in this area, another one could show all the utilities in that area and still another the buildings with their addresses etc, etc. In some cases property lines and/or contours are superimposed in white ink on the photo itself. By having transparent sheets for each category of information, any combination can be printed according to the requirements of an agency or individual.

With the use of an automatic computer-controlled drafting system, map information can be inputed by punched tape or cards and even stored so again various combinations may be reproduced. In other words, you can also have a computerized data bank of certain types of overlay material. This process is especially suited to cadastral maps developed in city engineer offices, county surveyor offices, assessor offices and the like. A note of warning, however, is in order here; do NOT assume the coordinates used for the development of this type of map to be acceptable for the determination or control of any legal property corners or lines.

Boundary

There must be finite lines of boundary by which, and within which, the information about the physical things in the field can be related on the map. This may be the outline of a lot on an already recorded map, a part of a section shown on a township plat or an irregular shape by a metes and bounds description as discussed in Chapter 3. Whichever it is, you should have this prepared and ready to receive the work done by the field crew. Having set the outside control for the map, the next step is to establish lines of traverse with station points in and around the area under survey.

Interior Control Traverse Lines

There are several methods available for an open traverse operation and the field man's decision as to which one to use depends on the terrain and the extent of coverage needed. One form sets up a center line with stationing marked at 25, 50 or 100-foot (10, 20 or 30 meter) intervals. Rod readings are taken not only at those station points but also at offsets to the right and to the left, an example of which is shown in Figure 1-9. If there is no map reference nor sketch attached, it is preferable to designate the offsets by compass direction from the center line. Each reading is taken at any distinct change in elevation. Either the rod reading, or the elevation reduced from it, is recorded in the field notes *above* the line and the distance from the center line is recorded *below* the line as shown in Figure 1-9.

An alternative of this method is the multi-control line pattern which, instead of using the single or center line with offsets, uses two or more lines. These may be arranged either parallel, in a fan shape or in a hodge-podge form, whichever way accommodates the lay of the land best. See Figure 6-1.

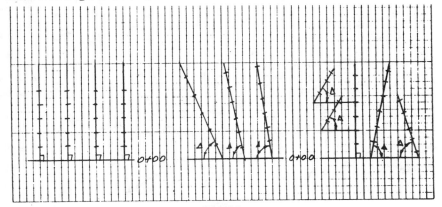

Fig. 6-1

It is of course necessary that every 0 + 00 station be oriented by angle *and* distance, either from the exterior boundary or a base line control. The rod readings are taken at every change of ground condition along each of the lines established.

An example of taking topography by the fan shape method is shown here where Figure 6-2 shows the survey of the lot lines for outside control and Figure 6-3 shows radial lines A, B, C and D set from the center of the road circle with ties along the north property line for control of the rod readings.

In some cases, it is preferable to just start from a corner of the project, or a tie point on the boundary, and from it establish a few interior stations, either on one straight line as shown in Figure 1-20 or a meander type line as shown in Figure 6-4.

The closed traverse method is a series of stations connected by successive distance and angle measurements around the circuit returning to the beginning station with one or more distance and angle ties to the exterior lines of the project. One example is shown in Figure 1-21. The

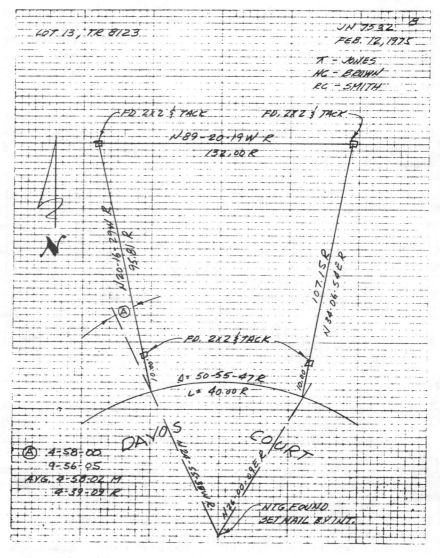

Fig. 6-2

closed traverse is preferable because you do not have points that may have gotten out of their true position due to an inadvertent mistake in measurement as might happen with the open traverse. In Figure 6-4, if Station 4 were tied to either Station 1 or the southeast corner of the property, you would have a good positional check as well as a mathematical check on the accuracy of the other stations.

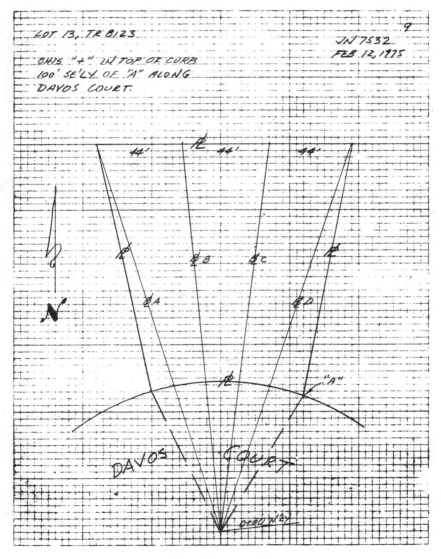

Fig. 6-3

Station Observations

After having established a station, there are two kinds of information to be acquired through it: (1) the horizontal location of elevation points and any pertinent objects within its purview and (2) elevations of the ground and/or objects around it.

There are three methods by which the horizontal position of points may be determined: (1) by distance and angle from an established station, the distance being measured either by tape, by reading a stadia rod through the transit or by an electronic measuring device; (Figure 1-26), (2) at the intersection of two lines from angles read at the opposite ends of a measured line (Figure 1-1), or (3) by reading two angles between three lines from one point to three established points: this last one being known as the 'three point problem' is more commonly used with a boat and sextant for the location of soundings (Figure 1-28).

The elevations are obtained either directly by reading the vertical height on a rod held at various points on the ground with the telescope of the level or transit set in the horizontal position and subtracting the reading from the height of the instrument (HI), or by mathematical deduction from reading a vertical angle to the rod and the distance on the rod between the upper and lower cross hairs in the transit telescope. The slope reading is then reduced by a sin-cos formula to the horizontal distance and vertical differential.

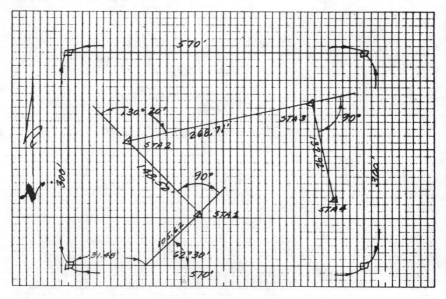

Meander type interior traverse.
Fig. 6-4

A much used method of locating objects is known as "stadia" which is based on the relationship of proportional triangles. When the rod is placed at the object (either in the center or at specific corners), the measurement on the rod × 100 is the distance as read between the upper and lower cross hairs in the telescope when the constant is 100, or, if the constant is different, whatever it is becomes the multiplier. See Figure 6-5. This may also be observed and reduced from a vertical angle reading to the rod in the same manner as for the elevations as previously described.

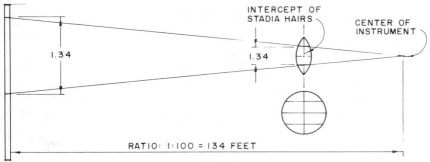

Fig. 6-5

In the foregoing situations, the positional direction of the elevation point or object from the station is established by reading the horizontal azimuth from a zero point as recorded in the field notes. In rare instances, compass readings may be used for positional direction.

Plotting

When the field work of the survey crew is turned in to the office, the draftsman takes the field notes and carefully plots all of the information on paper. As previously stated, he should have already made preparation for this by drawing the boundaries of the project from the legal description of the property. This should now be checked against the field work to determine if any monuments were found that should be shown on the map and if any of the field measurements are different from those given in the deed.

Where courses in the description tie to some definite thing, such as a street, a boundary deed, a stone, an iron pipe, etc., then the distance measured by the survey crew should be used either in addition to, or in lieu of, that given in the deed. If both are to be shown, the one taken from the deed is usually set in parenthesis following the measured amount, with the letter "D" for deed or the letter "R" for record denoting the fact that it is a matter of public record thus: 264.76 ft. (265.06 D) or (265.06 R). Sometimes the first figure is followed by "M" to denote "Measured" but it is not really necessary when used in conjunction with the other designated figure.

Having established and verified the perimeter within which the project is to be drawn, it is now ready to receive the other part of the field work. Along the Base Lines ₵ A, B, C and D shown in Figure 6-3, the following Field Notes were taken. (Figures 6-6 and 6-7) Along ₵ "A"

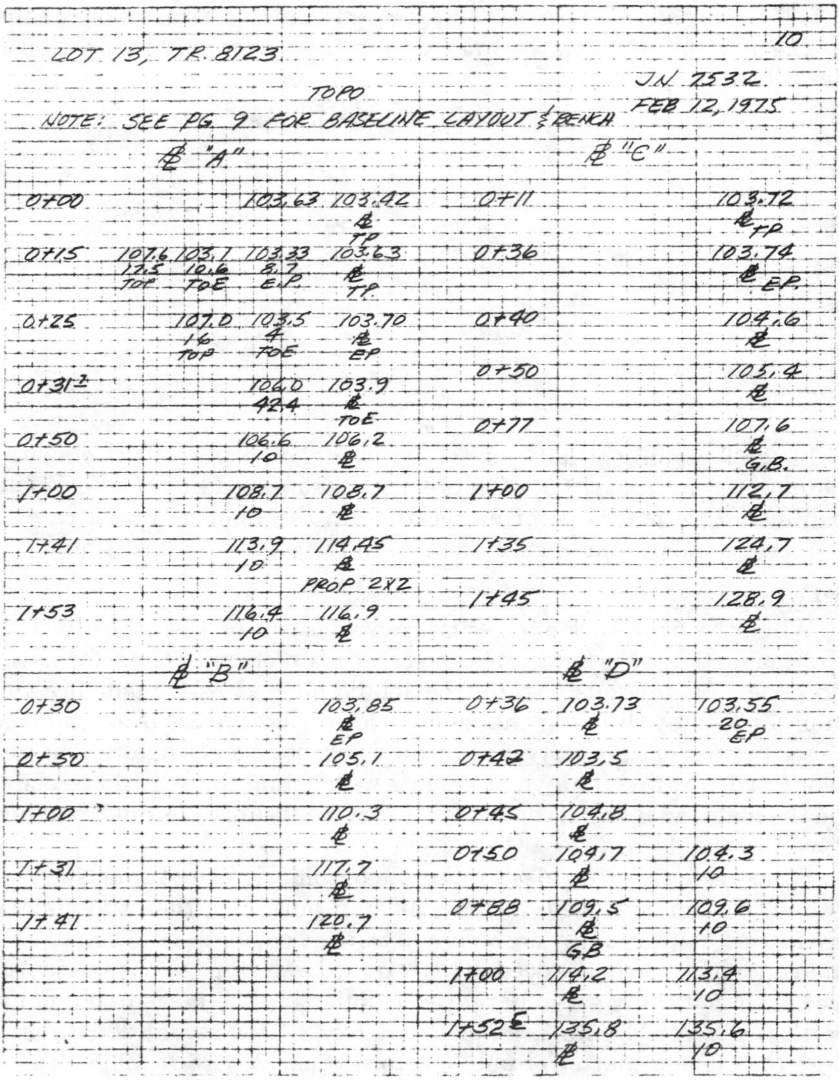

LOT. 13, TR. 8123.   10.

                              TOPO   J.N. 7532.
NOTE: SEE PG. 9 FOR BASELINE LAYOUT & BENCH   FEB. 12, 1975.

                    ₵ "A"   ₵ "C"

0+00   103.63 103.42   0+11   103.72
                              ₵   ₵ TP
                              TP
0+15   107.6 103.1 103.33 103.63   0+36   103.74
       12.5 10.6   8.7   ₵   ₵ EP
       TOP TOE   E.P.   TP
0+25   107.0 103.5 103.70   0+40   104.6
            16   4   ₵   ₵
            TOP   TOE   EP
0+31½   106.0 103.9   0+50   105.4
                   42.4   ₵   ₵
                           TOE
0+50   106.6 106.2   0+77   107.6
                          10   ₵
                                                              G.B.
1+00   108.7 108.7   1+00   112.7
                          10   ₵
1+41   113.9 114.45   1+35   124.7
                          10   ₵   ₵
                         PROP 2X2
1+53   116.4 116.9   1+45   128.9
                          10   ₵   ₵

              ₵ "B"   ₵ "D"

0+30   103.85   0+36 103.73   103.55
                          ₵   20
                          EP   EP
2+50   105.1   0+42 103.5
                          ₵   ₵
1+00   110.3   0+45 104.8
                          ₵   ₵
1+31   117.7   0+50 104.7   104.3
                          ₵   ₵   10
1+41   120.7   0+88 109.5   109.6
                          ₵   ₵   10
                                      G.B.
                                1+00 114.2   113.4
                                       ₵   10
                                1+52 E 135.8   135.6
                                       ₵   10

Fig. 6-6

you see that readings were taken not only on the line going uphill but also at certain offset points to the left of that line. While there were no offset readings from ℄ "B" and "C", some were made to the right from ℄ "D." In addition to the elevations recorded on the first sheet, a

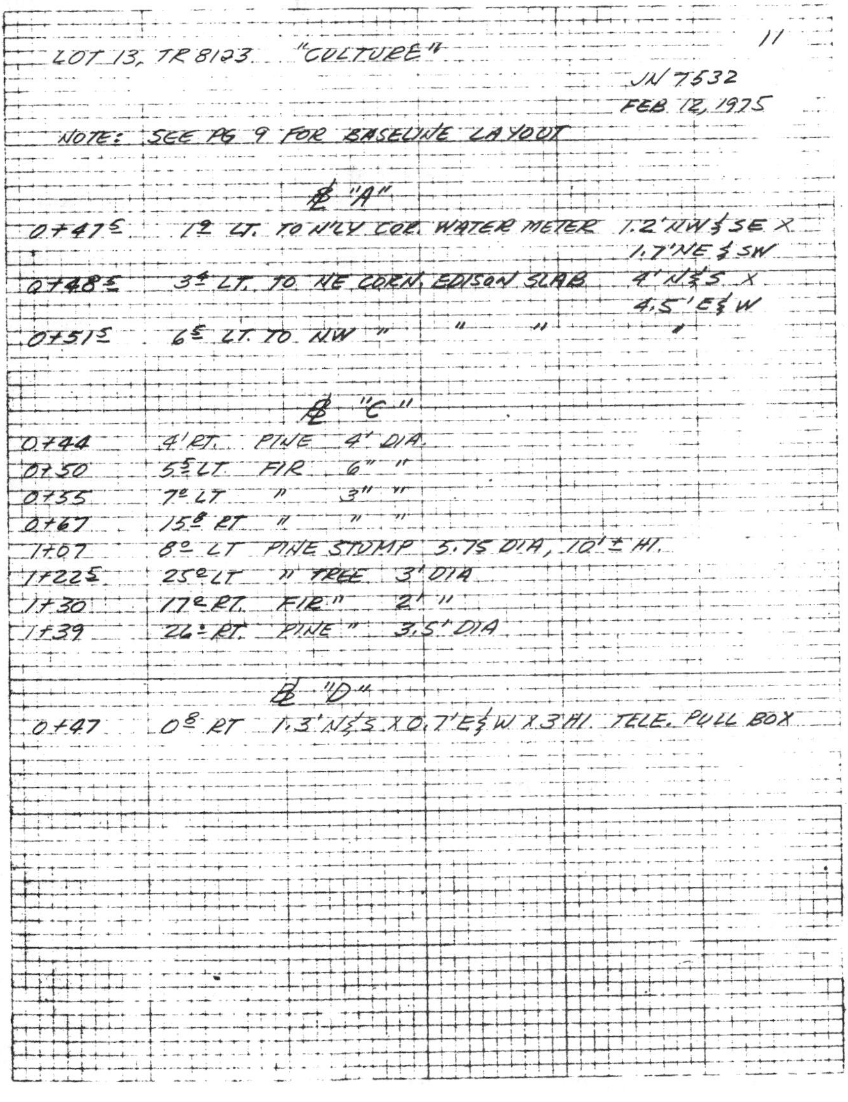
LOT 13, TR 8123. "CULTURE" 11
JN 7532
FEB 12, 1975
NOTE: SEE PG 9 FOR BASELINE LAYOUT
℄ "A"
0+47± 7± LT. TO N'LY COR. WATER METER 1.2' NW & SE X 1.7' NE & SW
0+48± 3± LT. TO NE CORN. EDISON SLAB 4' N & S X 4.5' E & W
0+51± 6± LT. TO NW " " " "
℄ "C"
0+44 4' RT. PINE 4" DIA.
0+50 5± LT. FIR 6" "
0+55 7± LT " 3" "
0+67 15± RT " " "
1+07 8± LT PINE STUMP 5.75 DIA, 10'± HT.
1+22± 25± LT " TREE 3' DIA
1+30 17± RT. FIR " 2' "
1+39 22± RT. PINE " 3.5' DIA
℄ "D"
0+47 0± RT 1.3' N & S X 0.7' E & W X 3 HI. TELE. PULL BOX

Fig. 6-7

separate group of notes (Figure 6-7) was made for the location of trees, water meter, etc. As shown in some of the examples in Chapter 1, all of the items, elevations and objects were mixed together in the field notes.

When a monument is found by the field crew, show it and its relationship to the particular line and/or corner of your deed description to which it applies. If several monuments are found, show *all* of them and the relationship of each to the other as well as to the line and/or corner of the property. The following example illustrates why you need to show all monuments for your own protection.

An architect made a request for the surveyor to set the four corners of his client's property and furnish a map showing the dimensions. Having acquired said map, the architect proceeded with his plans. Making maximum use of the available area, he designed the buildings to be set only two inches from the property lines. After the contractor completed the structures, the adjoining owner decided to also build and ordered a survey of his property. The map submitted by his surveyor showed that the recently erected building encroached onto the rear of his land almost four inches. Inquiry to the contractor and surveyor of the first property revealed the following facts:

1. The surveyor found three monuments at the rear (NE) corner and, in addition, set his own iron pipe.

2. The surveyor showed only his own monument at said rear corner.

3. The contractor saw only one iron pipe at said rear corner and he ran his string line from it to the iron pipe at the front (SE) corner. He did not dig to see if there were any other monuments.

4. The iron pipe used by the contractor was about six inches east of the one set by the surveyor, thereby putting the building four inches over the line.

Since a contractor is not supposed to be a surveyor (and dig for monuments), it is up to the surveyor to make his points definite, both on the ground, on the map and in the description.

Elevation Interpolations

To give you a better picture of what the field man must look at and the pertinent points which he must record so as to give you, the draftsman, the correct concept of the actual relief on the ground, look at Figure 6-8 and note the different profiles. The black dots on the Plan indicate the points where the gradient of the slope changes.

After plotting all of the objects and points of elevation on the map, the next step is to establish the lines of contour which are to be shown. First,

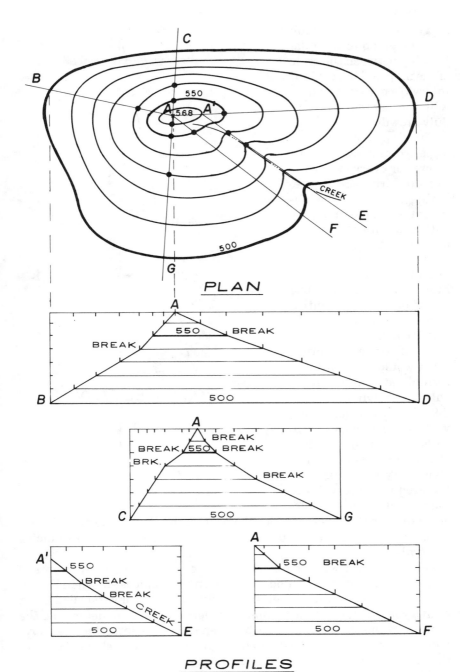

PLAN

PROFILES

Fig. 6-8

however, consideration
must be given to the relationship between the scale
of the map and the contour
interval so that nowhere
will the contours be un-
duly crowded.

If the land has very little
slope, contours should be
shown for every foot while
terrain with some appre-
ciable slope can be de-
lineated satisfactorily
with contours at two-foot
or five-foot intervals. In
metric we would consider
one-meter or two-meter
intervals. When the slope
is very steep such as with a
vertical difference of fifty
feet in a horizontal dis-
tance of ten feet, contour
lines would be more dis-
cernible at ten-foot intervals as shown in Figure
6-9. If this condition, along
with steeper slopes, pre-
vails over a major portion
of the area, then twenty-
five, fifty or even one-
hundred-foot interval contours would be preferable,

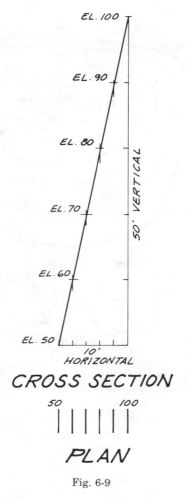

Fig. 6-9

depending on their closeness developed from the scale of the map. Again,
in metric, we would talk about ten, twenty, thirty or even fifty meter
intervals.

Let us proceed with the method of interpolating between points of
elevation taken on the ground. A beginning example is shown in Figure
6-10 with two points at an even six-foot differential. Elevation 642.0 feet
minus Elevation 636.0 feet equals six feet; therefore, to determine the
position of the one-foot contours between the two points, that measurement on your map is to be divided into six units and each one becomes
the control point for its respective contour line in that area.

If the reported elevation points are not at even feet, the same procedure is used but taking into consideration the fractional amount to be

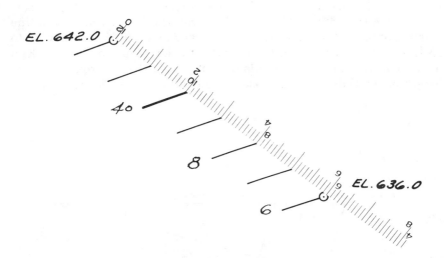

Fig. 6-10

distributed. For instance, between elevations 631.6 and 618.2, you have a vertical difference of 13.4 feet; therefore, in using one-foot contours, you want to determine the position of the contour levels 619 through 631 between these two points on your map. To set these points, find the particular scale on any ruler, whether it be an engineer's, an architect's or a metric scale, which will match 13.4 units between these two points. By setting the figures, that is (6) 18.2 and (6) 31.6 of your scale on their corresponding points on the map, you can then mark the one-foot intervals between them as shown in Figure 6-11. In this way, the 619 contour will be 0.8 of a foot above the El. 618.2 and the 631 contour will be 0.6 of a foot below the El. 631.6

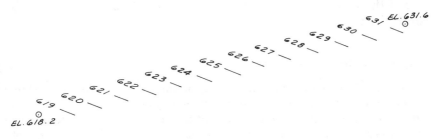

Fig. 6-11

There are instances where the divisions on a scale just do not fit the interval on your drawing. Interpolating for the position of the contour lines in such a case calls for the development and use of the proportionate triangle theory. Using the interval 317.6 to 327.4, draw a line starting at one of the reported elevations and at a small angle (say 15° to 25°) away from the true line to the next reported elevation, then mark on it the required contour intervals from *any* scale that approximates the length of the line between those two elevations. See Figure 6-12.

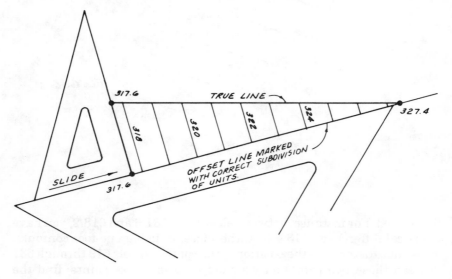

Fig. 6-12

Place the edge of a small triangle on a line between the end of the *marked* line, which diverges from the true line, to the corresponding elevation point on the true line and butt the long edge of another triangle up to it. Then slide the small triangle toward the coverging point of the two lines and transfer the contours from the *marked* line over to the true line.

Contour Delineation

A contour line represents the tracing of one constant elevation above or below a certain datum. In other words, every point on any one contour line is at the same elevation throughout that contour line. Although the datum may be an assumed point of elevation accommodating to the job, it is more often based upon Mean Sea Level. The determination of this control is developed from an average of the high and low tides as recorded on tide guages.

Bench Marks which designate certain elevations are imprinted brass caps set on or in a base of stable material and are established by survey crews.

The vertical difference is measured from the Mean Sea Level of a certain control area to the position of the Bench Mark and that elevation is stamped into the brass cap.

Contours are shown on a map at equally spaced *vertical* (not slope) measurements and the vertical distance between them is known as a contour interval.

In Figures 6-9 through 6-12, you were shown how even foot elevation points were established between the elevation points taken and recorded by the survey crew in the field. The next step is to draw lines through those points in such a manner as to represent the ground condition as close as possible. Each interpolated point of elevation must be joined to its corresponding point in the group of interpolated points on each side of it. The lines joining the corresponding elevation points must not have angles at the points as indicated in Figure 6-13 (unless the field notes expressly call for it) but should be curved as you would expect the slope of a hillside to curve as in Figure 6-14.

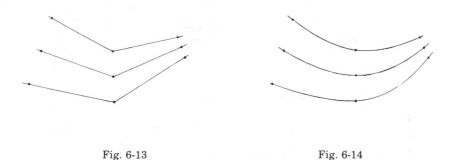

Fig. 6-13   Fig. 6-14

The three drawings in Figure 6-15 show contour lines in a perspective view, in a planimetric form and on a cross-section.

Note that a steep slope is represented by closely spaced contours, a gentle slope by comparatively widely spaced contours and a uniform slope, whether it is steep or gentle, is recognized by equally spaced contours.

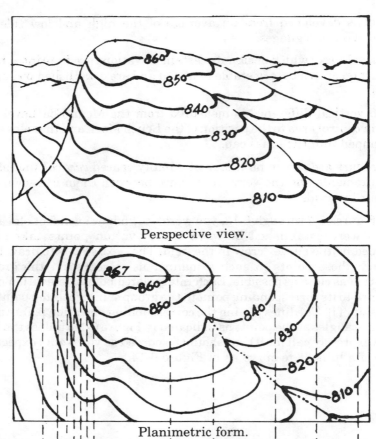

Perspective view.

Planimetric form.

Cross-section.

Fig. 6-15.

EXERCISE 6-1: On the drawing below, the interpolation has already
been made from the field shots and the even 10 foot interval
points are given. You are to connect the identical elevation
points with smoothly curved contour lines except at the crossing
of the creek where they should have a U or V form with the
curve or point heading upstream.

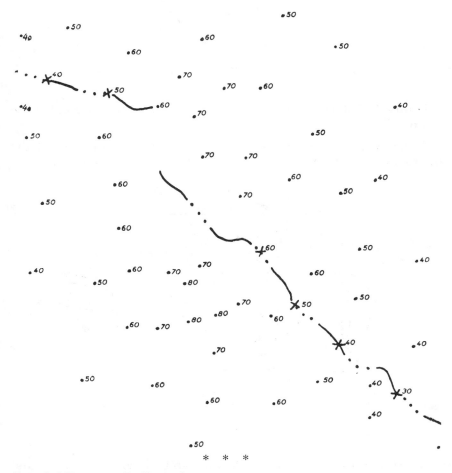

* * *

Special Feature Delineations

Besides the elevations recorded in the field notes, there are usually
side notes and sketches included to give the draftsman more complete
directions in delineating the true picture. For example, the field note
sketch shown in Figure 6-16 gives added details of the mountain ridge,
saddle, draw, creek line and spur line. The arrows indicate the direction
of water flow.

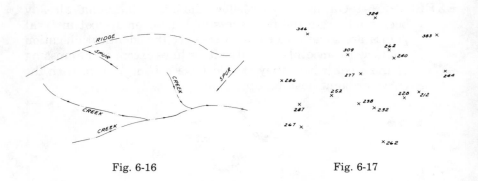

Fig. 6-16   Fig. 6-17

Figure 6-17 shows critical points of elevation for development of the contour lines.

Figure 6-18 shows the development of the contours based on both the interval derivation and the notations of the ground conditions.

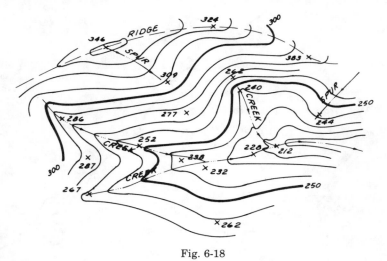

Fig. 6-18

Note that the configuration of the lines around the nose of ridges and spurs is generally of a "U" form while those that converge on a creek are closer to a "V" form. The sharpness of the "V" depends on the width of the stream bed; the wider the bed is, the less sharp are the contours.

Various features shown by the perspective and plan drawing in Figure 6-19 have been labeled to help you relate the method of indicating them on a flat surface and conversely to recognize their meaning when you look at an already prepared map.

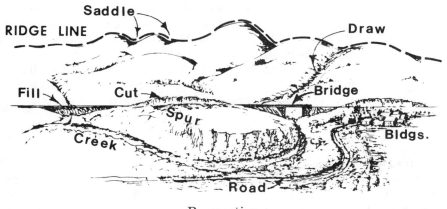

Perspective

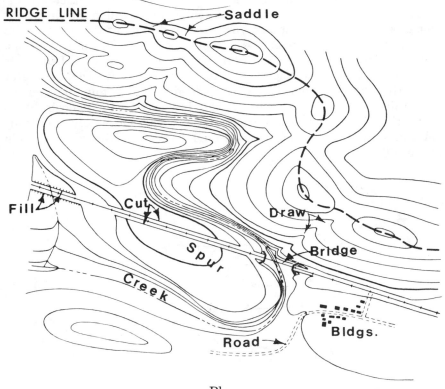

Plan

Fig. 6-19

EXERCISE 6-2: Take the field notes shown in Figures 6-2, 6-3, 6-6 and 6-7, lay out a map of the Lot 13 so surveyed with the radial control lines A, B, C, D and plot the elevations and topography recorded in the notes. Follow this with developing the contours. If drawn at a scale of ⅛″ = 1′, it will fit on a sheet 24″ x 18″ but if you prefer to make it twice as large on a ¼″ = 1′ (1″ = 4′) scale, you may. This is the type of map required by architects.

*   *   *

Slope Delineations

There are several methods employed to graphically illustrate the slopes of hills on a flat paper, namely: countour lines, horizontal shading, vertical shading and oblique shading. On some maps this is done in color, especially on U.S.G.S. and U.S. Army quadrangle sheets.

As you found from the previous discussion, contour lines are developed by interpolation between critical points of which the elevations have been recorded. Just as arrows are used to point the downstream direction of a creek or river, so are they used for the downhill slope of a draw, hillside or spur. In addition, when one or more closed contours represent a depression below the surrounding area, with a depth greater than the contour interval, ticks are added to the contour lines on the downhill side as shown in Figure 6-20. In such cases, it is desirable to show the elevation of the lowest depression point. Obversely, it is desirable to show the elevation of maximum height of the hilltop as shown in Figure 6-8.

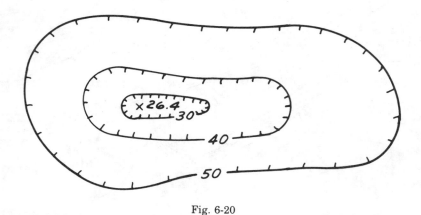

Fig. 6-20

A contour never splits nor do two contour lines join into one as illustrated in Figure 6-21

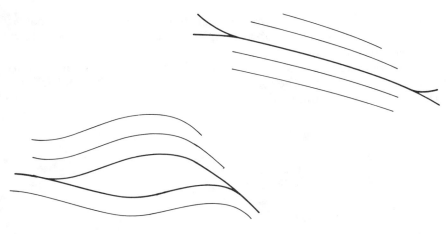

Fig. 6-21

Normally, one contour is not shown crossing another but in the situation where there is an overhanging cliff with an undercut, it may be necessary in order to clarify the picture, in which case the lower elevation contours should be dashed as shown in Figure 6-22

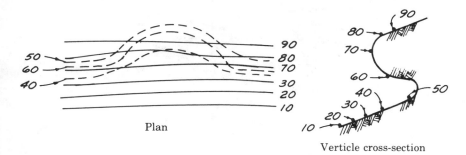

Plan

Verticle cross-section

Fig. 6-22

To further assist in reading the map certain interval contours are drawn with a heavier line to set them apart from the others; these are called *index* contours. Again look at Figure 6-8. If one-foot intervals are being used, the five-foot contours would become *index*. With two-foot intervals, the ten-foot contours would normally be drawn heavier as index lines. With small scale maps, the index lines might be the 100-foot or 500-foot contours. In metric, you would probably use index lines at 3-meter, 10-meter, 20-meter, 30-meter or 100-meter intervals.

Although contour lines give the picture of definite elevations, other methods are used to illustrate hill slopes in a merely graphic form. For example, lines may be drawn representing contour forms but not at specified elevations, as illustrated in Figure 6-23. This style is known as horizontal shading.

Another method known as vertical shading is the drawing of short lines at right angles to the contour lines as shown in Figure 6-24. After the short lines are inked, the contour lines are erased.

 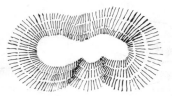

Fig. 6-23   Fig. 6-24

Oblique shading is a form of drawing in perspective with accent on the shadows as shown on the eroded bluff in Figure 6-25.

Fig. 6-25

EXERCISE 6-3: Take the set of annotated field notes in Figure 1-32 for Lot 15 in BIRCH WOODS Tract and draw a complete topo map suitable for the architect to use in developing the site and planning a new home for Mr. & Mrs. Sims. Start with outline of the lot, the half width street and the private driveway first marking the stations for reference. Do your preliminary work on a hard copy or work sheet plotting all of the subject matter and points with their elevations and any descriptions such as '4" orange'. After interpolating for the contours and drawing them, trace the final map and leave off all the scratch lines and calculations. This will make a good "show" map for you to use as an exhibit of your talent.

* * *

Scribe Coat Process.

With the increased use of photo processes in recent years, new techniques have been developed one of which is drawing with a metal scriber cutting through the coating on a film processed sheet. This makes a negative drawing as shown in Figure 6-26. From this a print can be produced in the positive attitude as in Figure 6-27.

The scriber used on film work is similar to the one shown in Figure 5-13 but instead of the pen point, it holds a steel tool which is available in various sizes to produce definite width lines. The scribing is usually done over a light table.

The background material from which scribing is developed may be either pen or pencil line work or photographic. For example, contours are extracted from the stereo model through the plotter onto a pencil manuscript map sheet; after the total sheet layout for the job is made, then the scribing is traced from them. At other times, the topography (all natural features) and/or the planimetry (all man-made features) will be traced directly from the transparent photos.

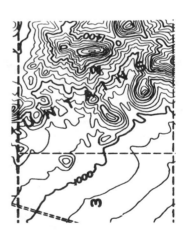

Fig. 6-26   Fig. 6-27

There are several advantages in using the scribing process. Corrections are easy to make with no tell-tale evidence because new coating can be brushed onto the film and new scribing done over it. The scriber renders an even width line at all times in contrast with ink lines which can vary.

It is also possible to print off of other drawings onto a scribe coat and add or subtract any desired information. Furthermore, because scribing is in negative form (Fig. 6-26), it requires fewer reproduction steps to obtain the finished map.

— NOTES —

Chapter *7*

Plan and Profile Drawings

Plan and Profile Drawings

These drawings are made for the pictorial designation of details concerning the construction of improvements. The "plan" is the horizontal layout of the project while the "profile" is the vertical cross section of that project. In the case of a pipeline, there will probably be need for a profile in only one vertical plane — that of the position of the pipe. For the construction of a roadway with curb, gutter and sidewalk, three profiles are usually needed — the centerline and the line of each curb. In the case of a country road with no curb or sidewalk, the three profiles would be centerline and either the edge of pavement or flow line if there is a gutter depression outside the pavement.

The existing ground profiles may be taken either directly from the field notes or interpolated from the contour map, the former being more accurate.

The relationship between the final grade shown by the designed line and the ground line elevations illustrates the amount of cut, or fill, involved in the project.

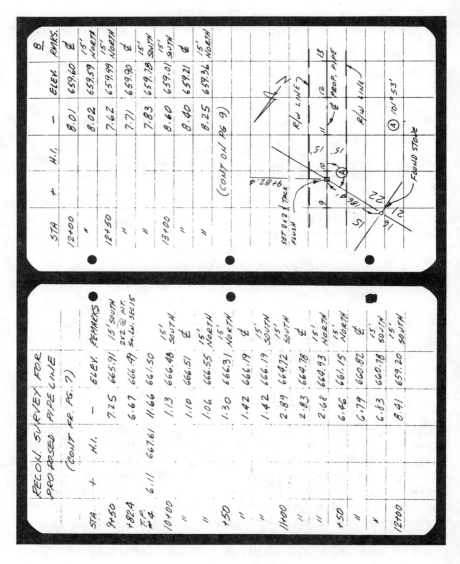

Fig. 7-1

Reduction from Field Notes

To begin with a simple project, we will develop a single line plan with profile for a pipe line. A short section is shown by the field notes in Figure 7-1 and illustrated in Figure 7-2.

Referring back to Figure 1-8, you see field notes of a single line profile for a sewer pipe line to be laid in the street and the amounts of cut and fill have been calculated and given to the field crew to mark on the stakes.

The amount of information required from the field crew depends upon the terrain and complexity of the topography involved. If the slope of ground on each side of the line is enough to require consideration of how the equipment can operate and where the diggings will be stored during the time the pipe is being laid, then sufficient additional lateral field notes will be needed for drawing a more comprehensive plan and profile.

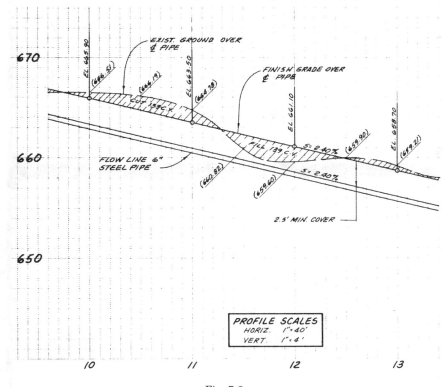

Fig. 7-2

Figure 1-9 shows a centerline run with "out" points for contour determination. Figure 7-3 here shows field notes of a Reconnaissance ('R') line with "out" points to cover a problem area as mentioned. Figure 7-4 shows the plan and profile developed from those field notes.

7.4

Sometimes a preliminary route study is needed to determine its feasibility before deciding on the right-of-way location. For this purpose, a profile line may be "picked off" from contours on U.S.G.S. or U. S. Army quadrangle sheets covering the area under consideration. Figure 7-5 shows a proposed line marked on a section of a quadrangle sheet and the profile immediately below it.

RECONNAISSANCE SURVEY (CONT FROM PG 15)

STA.	REMARKS	ELEV.	—	H.I.
18+00	'R' LINE	701.26	+ 5.96	707.22
"	10' NORTH	702.62	4.60	
"	20' SOUTH	699.42	7.80	
18+25	20' SOUTH	700.12	7.10	
"	'R' LINE	701.41	5.81	
"	10' NORTH	701.98	5.24	
18+50	20' NORTH	703.16	4.06	
"	20' SOUTH	701.12	6.10	
18+85	4' SOUTH 12" PINE	699.83	7.39	
19+00	20' SOUTH	700.01	7.21	
"	'R' LINE	698.83	8.39	
"	10' NORTH	703.76	3.46	
+25	10' NORTH	704.12	3.10	
"	'R' LINE	698.43	8.79	
"	20' SOUTH	700.36	6.86	
19+50	20' SOUTH	700.52	6.70	

STA.	REMARKS	ELEV.	—	H.I. 707.22
19+50	'R' LINE	699.20	8.02	707.22
"	10' NORTH	705.03	2.19	
19+65	5' NORTH 14" HEMLOCK			
19+75	10' NORTH	704.68	2.54	
"	'R' LINE	699.97	7.25	
"	20' SOUTH	700.76	6.46	
20+00	20' SOUTH	701.13	6.09	
"	'R' LINE	699.30	7.92	
"	10' NORTH	703.21	4.01	
+25	10' NORTH	702.76	4.46	
"	'R' LINE	699.46	7.76	
"	20' SOUTH	702.03	5.19	
+50	20' SOUTH	702.15	5.07	
"	'R' LINE	700.18	7.04	
"	10' NORTH	702.12	5.10	
20+71	15' SOUTH 24" PINE			
20+75	20' SOUTH	701.82	5.70	
"	'R' LINE	700.62	6.60	

(CONT. ON PG. 17.)

Fig. 7-3

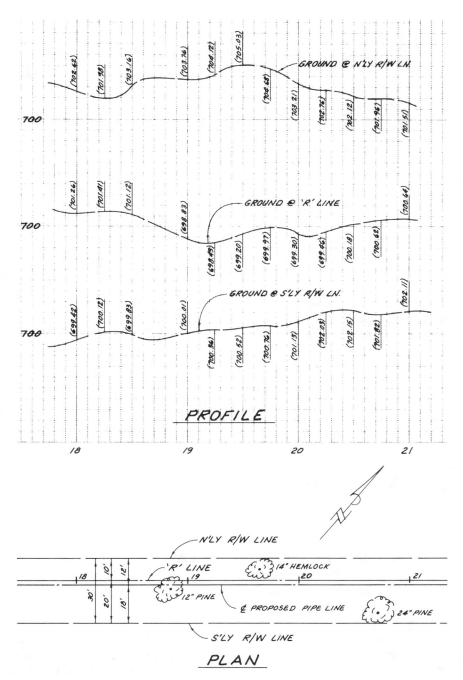

PROFILE

PLAN

Fig. 7-4

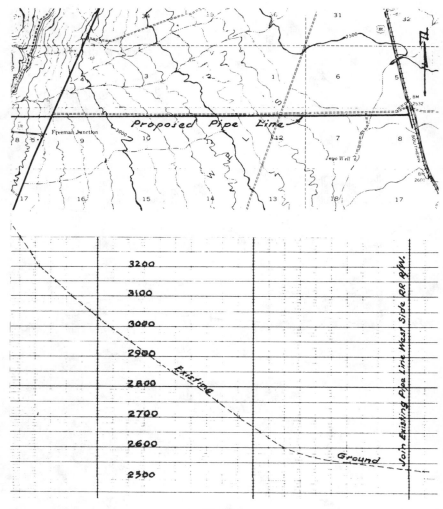

Fig. 7-5

Development of Final Drawings for a Street Improvement

We will now consider the actual work to be done in preparation for the final drawings used for construction of the improvements.

Format

Either paper or cloth material is available preprinted with the profile format with vertical lines marked for every ten feet at a horizontal scale of 1″ = 40′ and the horizontal lines drawn in groups of five for use at a vertical scale of 1″ = 4′ or multiples thereof. The whole sheet may be printed in this form but usually one-half the width is left blank for the "plan" which may be either the upper part or the lower part.

There are generally two ways of presenting such plans: 1 - a series of sheets cut to a set standard length, or 2 - one continuous sheet for the entire length of the project if it is not too long. Also you will notice that some sets of drawings show the plan on top while others place the profile above the plan. This alternative of arrangements is really immaterial and is usually dictated by some internal policy or individual preference.

Where standard size sheets are used, the company often has them printed with its name block, including line for title and signature of private engineer, blocked-out spaces for sheet number, job number and project, identification, lines for date of original drawing plus dates of revisions, and lines for title and signature of public engineer approving plans.

Requirements

In preparing street improvement plans, the amount of detail is determined by the prevailing conditions. If it is in a city with underground pipe lines, overhead power on poles, storm drains, etc., there will be much detail to consider. If, on the other hand, the project is in the country without those problems, the plans will be simpler.

Regardless of the amount of improvements, each set of plans requires specifications and information concerning details of construction on that job. These items are usually listed under such headings as GENERAL NOTES, CONSTRUCTION NOTES, ESTIMATED QUANTITIES, NOTICE TO CONTRACTORS, etc., all of which are placed on the first sheet, also known as the title sheet, and carried over to the second sheet if necessary.

Another item which should be on the first sheet is the Key Map, also known as Location Map, showing the location of the project in relation to area land marks such as streets, highways, railroads, cities, etc. If there are many sheets to the job, an Index Map should also be included so that the sheet of any particular area can be quickly found.

The benchmark used to control the elevations on the project should be shown on the first sheet anyway and sometimes repeated on subsequent sheets.

Following are some of the items, although there can be more, or fewer, which belong under the heading of GENERAL NOTES (and not necessarily in the order given):

a - Basis of stationing (street intersection, etc.).

b - Basis of earth work computations (photogrammetric or otherwise).

c - Conformity with certain standard specifications.

d - Pavement and subgrade preparation dependent upon soil tests.

e - Preservation and/or replacement of survey monuments.

f - Protection of existing improvements.
g - Relocation of underground or overhead utilities.
h - Removal of trees, debris, etc.
i - Repairs and sealing of finish pavement.
j - Street sign schedule.
k - Tests required.
l - Trees.
m - Underground structures details.

Under the heading of CONSTRUCTION NOTES, you will have such items as:

a - Street.
    1 - Compaction.
    2 - Cross-gutters.
    3 - Curb and gutter.
    4 - Driveway aprons.
    5 - Guard rails.
    6 - Pavement.
    7 - Sidewalk.
    8 - Signs.
    9 - Survey monuments.
    10 - Water laterals and meters.
b - Sewer and/or storm drain.
    1 - Inlet structures.
    2 - Junction chambers.
    3 - Outlet structures.
    4 - Pipe.
c - Underground utilities (other than b).
    1 - Pipe.
    2 - Valves.
    3 - Vaults.

ESTIMATED QUANTITIES may include but are not limited to:
a - Excavation and deposit of on-site dirt.
b - Disposition of excess dirt.
c - Aggregate base.
d - Asphalt concrete.
e - Prime coat.
f - Seal coat.
g - Sidewalk.
h - Curb and gutter.
i - Guard rails.
j - Reflector paddles.
k - Sewer laterals.
l - Water laterals.
m - Trees.

Under NOTICE TO CONTRACTORS, such items as safety, liability, responsibility and performance are delineated as pertaining to the job.

Although the standard cross-sections of each type of road, curb, gutter, driveway and walk to be used on the project may be detailed on the first or second sheet, special details are often shown near the place in the project to which they apply. If the specifications include certain "Standards" established by the office in charge of public works, then any of the above items so established do not need to be duplicated.

To assist in cross referencing, numbers in circles are placed on the drawings at the pertinent points, or with leaders, and duplicated opposite the applicable item under the above headings.

Drawings - plan.

On the preprinted plan-profile sheet, the unruled area, which can be placed either at the top or bottom of the sheet as previously mentioned, is used to illustrate the plan layout of the project.

In the case of a subdivision, this drawing is usually copied from the respective streets shown on the tract map. If there happens to be two segments of street improvement, one on each side of an existing improved street, the existing work will be shown in dash lines so as to illustrate the relationship of the parts and with appropriate notes.

On a cross-country highway, pipe line or power line, the right-of-way, as described in deeds of record or as designed prior to acquisition, is drawn accordingly in the unruled area. On this type of project, sometimes the topography and contour lines in and around the right-of-way are included, either taken from the reconnaissance survey or from photogrammetry. In some cases, the right-of-way is drawn directly onto orthophotos and then reproduced in the blank area of the plan-profile sheet. Having this background information along the right-of-way is an excellent assist in determining the amount of slope easements that may be necessary and furthermore, they can be delineated on the plan at the same time.

Besides showing the pertinent mathematical data, the aforementioned cross reference numbers in circles are placed in their proper positions on the plan.

Drawings - profile.

As mentioned, you will probably have either a single line or triple line profile job. In either case, you should plot the *natural ground* elevations along the one line or the center line, keeping in the middle range between the top and bottom of the ruled area on the sheet. The points of elevation along the stations must agree exactly with the stationing on the plan and vice versa.

If the increase of elevation gets too high or too low, stop it and repeat the identical ground elevation at a higher or lower position on the vertical line of the same station as shown in Figure 7-6.

If you are to show the ground elevations of the side lines, they will be run on the same identical stations, one in the upper part and the other in the lower part of the ruled area. Be sure to label the elevation control lines with bold figures.

After the engineer designs the position of the pipe line or the finish grades for the center of pavement and top of curb, you draw them in

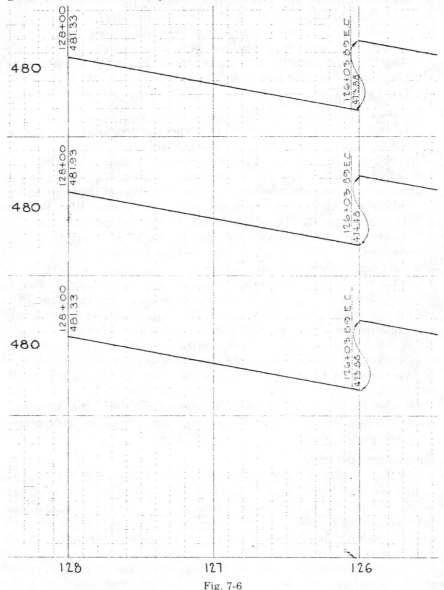

Fig. 7-6

relationship to the ground lines and designate the percentage of grade in between stations that control it and show both the elevation and station at each and every change of grade. Figure 7-7 illustrates a short section of plan and profile for a street improvement showing center line of pavement and top of curb on each side.

The next 6 pages show a complete PLAN & PROFILE set including Notes.

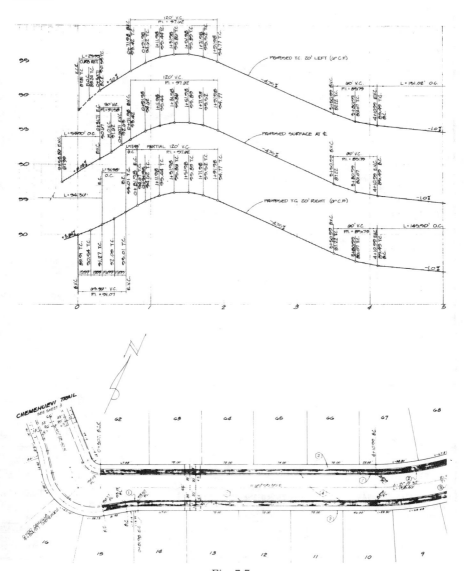

Fig. 7-7.

GENERAL NOTES

PLAN AND PROFILE FOR TH
IMPROVEMENT OF STREETS WIT

TRACT 1

STREET IMPROVEMENTS

1. THIS NOTE HEREIN INCORPORATES BY REFERENCE THOSE GENERAL NOTES NUMBERED 1 THROUGH 13 INCLUSIVE OF E M A STANDARD PLANS 801 1980 EDITION
2. ALL STREET STATIONING REFERS TO CENTERLINE OF STREET
3. ALL HIGHWAY SIGNS AND STREET NAME SIGNS SHOWN ON THE PLAN MUST BE SUPPLIED AND INSTALLED BY THE CONTRACTOR PER PLAN COUNTY ENVIRONMENTAL MANAGEMENT AGENCY STD PLANS NO 407 408 AND 409
4. ALL GRADE BREAKS IN RCP STORM DRAINS SHALL BE CONSTRUCTED USING EITHER BEVELED PIPE OR A CONCRETE COLLAR AT OPTION UNLESS OTHERWISE NOTED
5. SPECIAL COVER PIPE SHALL HAVE A MINIMUM OF 1 1/2 CLEAR COVER FROM THE INVIDE OF THE PIPE TO THE REINF.R NG STEEL
6. ON ARTERIAL HIGHWAYS THE HINGE POINT FOR ALL EXCAVATION OR EMBANKMENT SHALL BE A MINIMUM OF TWO FEET OUTSIDE RIGHT A-Y
7. ALL CONCRETE CURB AND GUTTER FLOW LINES WITH LESS THAN 1% GRADE SHALL BE WATER TESTED PRIOR TO FINAL FINISHING TO INSURE PROPER DRAINAGE WITHOUT UNACCEPTABLE HIGH OR LOW SPOTS
8. ALL UTILITY TRENCH BACKFILL AND COMPACTION INSPECTION OUTSIDE THE LIMITS OF DEDICATED STREET RIGHT OF WAY SHALL BE PERFORMED BY O C E M A REGULATION
9. ALL CONCRETE SIDEWALKS OR CURBS SHALL BE SAWCUT TO THE NEAREST TRANSVERSE SCORE MARK IF ACCEPTABLE OTHERWISE AT A WEAKENED PLANE JOINT AND REPLACED IN CONFORMANCE WITH THE APPLICABLE PROVISIONS OF THE ORANGE COUNTY STANDARD SPECS HIGHWAY PLANS AND STANDARD SPECIFICATIONS
10. DEVELOPER SHALL MAINTAIN ADJACENT STREETS IN A NEAT SAFE CLEAN AND SANITARY CONDITION AT ALL TIMES AND IF THE LAST DAYS OF COUNTYS INSPECTOR THE ADJACENT STREETS SHALL BE KEPT CLEAN OF DEBRIS WITH DUST AND DIRT NOT LAN BEING LABELLED ALL TIMES DEVELOPER SHALL BE RESPONSIBLE FOR ANY CLEAN UP ON ADJACENT STREETS AFFECTED BY HIS CONSTRUCTION METHODS STREET CLEANING SHALL BE BY DRY SWEEPING IF ALL PAVED AREAS
11. PRIOR TO FINAL ACCEPTANCE OF STREET IMPROVEMENTS ALL STREET STRIPING AND STENCILING AS THE PERIMETER OF THE LIMITS OF PROJECT WILL BE RESTORED TO A LIKE NEW CONDITION IN A MANNER MEETING THE APPROVAL OF THE ASSISTANT OF THE RANGE DEVELOPMENT
12. STENCIL CONTROL INFORMATION SHALL BE PER STANDARD SYMBOL LEGEND AS APPROVED EQUAL

SEWER SYSTEM

1. STATIONS SHOWN THUS ⬛ ARE SEWER STATIONS AND ARE INDEPENDENT OF STREET STATIONS
2. SEWER SYSTEM AS SHOWN ON THE PLANS SHALL BE CONSTRUCTED IN ACCORDANCE WITH THE STANDARD DRAWINGS AND SPECIFICATIONS OF THE SANTA MARGARITA WATER DISTRICT THE CONTRACTOR SHALL KEEP A COPY OF THE STANDARD SPECIFICATIONS IN THE JOB SITE AT ALL TIMES
3. TYPE I BEDDING WILL BE USED WHEN SUBSTITUTE SAND LINE INFERIOR IS NOT IN WARE ENCOUNTERED IN THE PIPE ZONE AS DETERMINED BY THE DISTRICT
4. FOUR COPIES OF THE CONSTRUCTION PLANS SHALL BE FURNISHED TO THE DISTRICT ENGINEER FOR HIS FINAL REQUEST FOR INSPECTION
5. THE DISTRICT ENGINEER SHALL BE NOTIFIED AT LEAST TWO WORKING DAYS PRIOR TO ANY INSPECTION ALL SEWER
6. ALL STUBS AND METERS JOINTS OUTFALL MANHOLES TO BE KEPT AT MAXIMUM MEASURED FROM THE INSIDE AL THE MANHOLE
7. SEWER RECTANGULAR MANHOLE PROFILE LINE HORIZONTAL STAKE LABELS BETWEEN ENTER NEW STAKE MANHOLES
8. SURFACE STAKE THE LOCATIONS OF ALL WATERLINES AND LATERALS NOT NORMAL OTHERWISE SEWER SHALL BE STAKED AT PROPERTY LINE AND TO EACH PROPERTY AND EASEMENT LINE
9. TO PREVENT AL CENTRAL USE OF THE NEW SEWER PIPE FOR COMPLETION AND ACCEPTANCE THE NEW PL OJECT CONNECTION MANHOLES SHALL BE SEALED WITH BRICKEN BRICK AND MORTAR IN STALLATION IN PLACE PLUG SHALL BE APPROVED BY THE ENGINEER PLUG SHALL BE REMOVED AT THE MEET FINAL INSPECTION
10. MANHOLE COVERS SHALL BE ADJUSTED AS NECESSARY AFTER STREETS ARE INSTALLED
11. WHERE THE REFERENCE ELEVATION BETWEEN THE INVERTS OF THE UPPER AND LOWER MANHOLES FOR A GIVEN MANHOLE EXCEEDS FEET THE HIGHER BETWEEN SHALL BE A DROP MANHOLE WHICH SERVICE THE NEW SEWER STATION AFTER
12. ALL HOUSE LEADERS PER LOT INCLUDING THE WET WELL ELEVATIONS OF SEWER THE MINIMUM LOT ELEVATION OF THE PER SEWER
13. PRIOR TO AND BEGIN AND BACKFILL INSPECTIONS OF THE SUBJECT OF THE FIRE LATERAL CALL SEWER

DOMESTIC WATER SYSTEM

1. THE WATER SYSTEM SHALL BE INSTALLED TO CONFIRM TO THE SANTA MARGARITA WATER DISTRICT STANDARD SPECIFICATIONS AT NO LEAST REVISED THE CONTRACTOR SHALL KEEP A COPY OF THE STANDARD SPECIFICATIONS IN THE JOB SITE AT ALL TIMES
2. FIRE HYDRANTS LINES SERIES STOCK OR APPROVED EQUAL SHALL BE INSTALLED PER SANTA MARGARITA WATER DISTRICT STANDARD SPECS
3. FIXATION ALL FIRE HYDRANTS SHALL HAVE TWO OUTLET AND A OUT OPENING NUT
4. THE DISTRICT INSPECTOR SHALL BE NOTIFIED AT LEAST TWO WORKING DAYS PRIOR TO STARTING CONSTRUCTION ON NEW SECTIONS
5. WATER LINES SHALL BE INSTALLED FEET FROM THE SURFACE UNLESS OTHERWISE SHOWN IN THE PLANS AND SUBSEQUENT TO THE INSTALLATION OF THE CURBS MINIMUM OF FEET VERTICAL AND FEET BELOW SUBGRADE
6. ALL WATER SERVICE SUPPLIED WITH PRESSURE HIGHER THAN 80 PSI SHALL BE PROVIDED WITH APPROVED PRESSURE REDUCING VALVES SET AT 65 PSI
7. THE DEVELOPER SHALL FURNISH THE SANTA MARGARITA WATER DISTRICT WITH EASEMENTS FOR THAT PORTION OF THE WATER SYSTEM OUTSIDE PUBL RIGHT OF WAY PRIOR TO COMPANY OF ACCEPTANCE OF THE WATER SYSTEM
8. FOUR COPIES OF THE CONSTRUCTION PLANS SHALL BE FURNISHED TO THE DISTRICT ENGINEERS FOR HIS REPORT REQUEST FOR INSPECTION
9. ALL FLANGED CONNECTIONS SHALL BE INSTALLED WITH OR WITHOUT FIELD BY A DIFFERENCE OF 24 HOURS AFTER INSTALLATION INCLUDING NUTS BOLTS AND FLANGES
10. N FA CHO TO BE BACKFILLED UNTIL INSPECTED BY THE SANTA MARGARITA WATER DISTRICT
11. SHUTDOWNS OF EXISTING LINES TO FACILITATE CONNECTION TO EXISTING FACILITIES SHALL BE COORDINATED WITH THE SANTA MARGARITA WATER DISTRICT
12. THE CONTRACTOR SHALL OBTAIN AN EXCAVATION PERMIT FROM THE COUNTY EMA PRIOR TO START OF CONSTRUCTION

RECLAIMED WATER SYSTEM

1. THE RULES ON AND INSTALLATION OF THE IRRIGATION WATER SYSTEM SHALL CONFIRM TO THE SANTA MARGARITA WATER DISTRICT STANDARD SPECS AT NO LEAST REVISED THE CONTRACTOR SHALL KEEP A COPY OF THE STANDARD SPECIFICATIONS ON THE JOB SITE AT ALL TIMES
2. AL WASTE IRRIGATION PIPING INSTALLED UNDER THIS CONTRACT SHALL BE IDENTIFIED AS RECLAIMED WATER PIPING IN CONFORMANCE WITH THE SANTA MARGARITA WATER DISTRICT'S REGULATIONS
3. AL EASEMENTS AND PIPE SHALL BE IN ACCORDANCE WITH THE SAME LATEST CUSTOM SPECS AS IT IN A ND IN ACCORDANCE WITH REGULATION
4. A MINIMUM OF PIPE SHALL BE AS REQUIRED BY STATE DEPARTMENT OF HEALTHS LATEST REQUIREMENTS FOR SEPARATION BETWEEN WATER MAINS AND SANITARY SEWERS
5. RECLAIMED WATER SYSTEM SHALL HAVE A MINIMUM ACI OVER TO BE CONCRETE ENCASED FOUR INCH PIPE SHALL BE MINIMUM CLASS 200
6. ALL VALVES SHALL BE BUTTERFLY VALVES WITH HI RESISTANT DISCS AND STAINLESS STEEL STEMS VALVE BOXES SHALL BE APPROVED BY MAG AND CENTIFIED AS OUTLINED IN THE DISTRICTS RECLAIMED WATER MANUAL
7. THE SYSTEM WILL BE INSPECTED AND PRESSURE TESTED IN ACCORDANCE WITH THE EMA RECLAIMED WATER SYSTEM SPECIFICATIONS

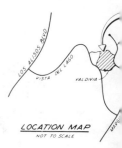

LOCATION MAP
NOT TO SCALE

LOS ALISOS BLVD
VISTA DEL LAGO
VALDIVIA

ALL UNDERGROUND UTILITIES OR STRUCTURES REPORTED BY THE OWNER OR OTHERS AND THOSE SHOWN ON THE RECORDS EXAMINED ARE INDICATED WITH THEIR APPROXIMATE LOCATION AND EXTENT THE OWNER BY ACCEPTING THESE PLANS OR PROCEEDING WITH IMPROVEMENTS PURSUANT THERETO AGREES TO ASSUME LIABILITY AND TO HOLD UNDERSIGNED HARMLESS FOR ANY DAMAGES RESULTING FROM THE EXISTENCE OF UNDERGROUND UTILITIES OR STRUCTURES NOT REPORTED TO THE UNDERSIGNED NOT INDICATED ON THE PUBLIC RECORDS EXAMINED LOCATED AT VARIANCE WITH THOSE REPORTED ON SHOWN ON RECORDS EXAMINED THE CONTRACTOR IS REQUIRED TO TAKE DUE PRECAUTIONARY MEASURES TO PROTECT THE UTILITIES OR STRUCTURES SHOWN AND ANY OTHER UTILITIES OR STRUCTURES FOUND AT THE SITE IT SHALL BE THE CONTRACTOR'S RESPONSIBILITY TO NOTIFY THE OWNERS OF THE UTILITIES OR STRUCTURES CONCERNED BEFORE STARTING WORK

CONTRACTOR AGREES THAT HE SHALL ASSUME SOLE AND COMPLETE RESPONSIBILITY FOR JOB SITE CONDITIONS DURING THE COURSE OF CONSTRUCTION OF THIS PROJECT INCLUDING SAFETY OF ALL PERSONS AND PROPERTY THAT THIS REQUIREMENT SHALL APPLY CONTINUOUSLY AND NOT BE LIMITED TO NORMAL WORKING HOURS AND THAT THE CONTRACTOR SHALL DEFEND INDEMNIFY AND HOLD THE OWNER THE CONSULTING ENGINEER AND THE COUNTY OF ORANGE HARMLESS FROM ANY AND ALL LIABILITY REAL OR ALLEGED IN CONNECTION WITH THE PERFORMANCE OF WORK ON THIS PROJECT EXCEPTING FOR LIABILITY ARISING FROM THE SOLE NEGLIGENCE OF THE OWNER THE CONSULTING ENGINEER OR THE COUNTY OF ORANGE

COUNTY OF ORANGE **E.M.A., REGULATION DIVISION**	**SANTA MARGARITA WATER DISTRICT**	**IMPROVEMENT DISTRICT NO. 1-W (WATER)**	**APPROVED FOR WATER DISTRIBUTION SYSTEM & FIRE PROTECTION FACILITIES**	**BASIS OF BEARIN**

THIS PLAN IS SIGNED BY EMA DEVELOPMENT FOR CONCEPT AND ADHERENCE TO COUNTY STANDARDS AND REQUIREMENTS ONLY EMA DEVELOPMENT IS NOT RESPONSIBLE FOR DESIGN ASSUMPTIONS AND ACCURACY

APPROVED _____ R.C.E. 0856 DATE 6-11-81

SANTA MARGARITA WATER DISTRICT — IMPROVEMENT DISTRICT NO. 1-S (SEWER) — APPROVED _____ DATE 5-11-81

SANTA MARGARITA WATER DISTRICT — IMPROVEMENT DISTRICT NO. 1-S (RECLAIMED) — APPROVED _____ DATE 5-11-81

APPROVED _____ DATE 5-11-81

APPROVED _____ DATE 4-2-81
COUNTY FIRE MARSHAL

THE BEARINGS SHOWN CENTERLINE OF ALA TRACT 9077 RECORD IV OF MISCELLANEOUS COUNTY OF CALIFORNIA

BENCH MARK EL
OF THE INTERSECTION
SOUTH FROM THE CENT
WEST FROM THE CENTER
SET IN THE TOP MON
UNDERGROUND CONCRETE

76

CONSTRUCTION NOTES & ESTIMATED QUANTITIES

CONST. NOTE NO.	ITEMS TO BE CONSTRUCTED OR INSTALLED	TRACT 11276	ALARCON	TOTAL	UNIT
	STREETS				
1	EXCAVATION (ROADWAY)	2769	59	2828	C.Y.
2	ASPHALT CONCRETE AGGREGATE BASE	35086	1596	36681	
	SC-250 PRIME COAT	1531	44	1575	
	SEAL COAT	611	18	629	
3	CURB AND GUTTER TYPE A-8 C.F. PER O.C.E.M.A. STD. PLAN N° 201				
4	CURB AND GUTTER TYPE A-2 -8 C.F. PER O.C.E.M.A. STD. PLAN N° 201				
5	CURB AND GUTTER TYPE D 6 C.F. PER O.C.E.M.A. STD. PLAN N° 201	2877		2877	
6	TRANSITION CURB AND GUTTER 6 C.F. TO 8 C.F.	73		73	
7	CURB ONLY AROUND RETURN				
8	CROSS GUTTER PER O.C.E.M.A. STD. PLAN N° 202	73		73	
10	CROSS GUTTER (4XL) PER O.C.E.M.A. STD. PLAN N° 208	880		880	
11	SIDEWALK PER O.C.E.M.A. STD. PLAN N° 207	8121		8121	
12	SIDEWALK ACCESS RAMP PER O.C.E.M.A. STD. PLAN N° 111	180		180	
13	DRIVEWAY APPROACH PER O.C.E.M.A. STD. PLAN N° 204	4195		4195	
14	STREET NAME SIGN PER O.C.E.M.A. STD. PLAN N° 405	2		2	
15	TIMBER BARRICADE PER O.C.E.M.A. STD. PLAN N° 404				
16	METAL BEAM GUARD RAIL PER O.C.E.M.A. STD. PLAN N° 403				
17	STREET LIGHTS PER O.C.E.M.A. STD. PLAN N° 411 (HPSV 5800) UNLESS SHOWN OTHERWISE	8	1	9	
18	REMOVE EXISTING BARRICADE				
19	MAILBOXES PER O.C.E.M.A. STD. PLAN N° 212	37		37	
20	6" A.C. DIKE PER O.C.E.M.A. STD. PLAN N° 201				
21	TEMPORARY 2" A.C. SIDEWALK				
22	TRAFFIC WARNING SIGNS	2		2	
23	TRAFFIC REGULATORY SIGNS				
24	STENCIL CONTROL INFORMATION				
25	12" SOLID WHITE STRIPE				
26	8" SOLID WHITE STRIPE				
27	4" X 100' SOLID WHITE STRIPE				
28	4" SOLID YELLOW STRIPE				
29	4" DASHED WHITE STRIPE PER DETAIL SHEET NO.				
30	4" SOLID DOUBLE YELLOW STRIPE	1520		1520	
31	60' PARABOLIC MEDIAN FLARE PER O.C.E.M.A. STD. PLAN N° 111				
32	90' MEDIAN TRANSITION PER O.C.E.M.A. STD. PLAN N° 111				
33	REMOVE 2" A.C. SIDEWALK				
34	REMOVE A.C. DIKE				
35	REMOVE A.C. PAVEMENT	81	1595	1676	
36	SLOPE TRAIL PER DETAIL SHEET NO.				
37	REMOVE CURB AND GUTTER	81		81	
38	REMOVE SIDEWALK	365		365	
	SEWER				
	8" P.V.C. SEWER MAIN	1435	125	1560	L.F.
	4" P.V.C. SEWER LATERAL	1172		1172	L.F.
	STANDARD MANHOLE	7	1	8	EA.
	CONCRETE ENCASEMENT				
	WATER (DOMESTIC)				
	8" CLASS-200 A.C.P. WATER MAIN ZONE	1042		1042	L.F.
	6" CLASS-200 A.C.P. WATER MAIN ZONE				
	4" CLASS-200 A.C.P. WATER MAIN ZONE	456		456	
	8" BUTTERFLY VALVE	3		3	
	8" DUCTILE IRON PIPE				
	FIRE HYDRANT ASSEMBLY	3		3	
	BLOW-OFF ASSEMBLY				
	AIR RELEASE VALVE				
	STANDARD 1" SERVICE	37		37	EA.
	HOT TAP				
	WATER (RECLAIMED)				
	4" CLASS-200 A.C.P. ZONE				L.F.
	4" CLASS-200 A.C.P. ZONE				
	2" METER AND SERVICE				
	2" METER				
	HOT TAP				
	STORM DRAIN				
	18" R.C.P. - 1500 D	72	156	228	L.F.
	CURB INLET L-21' TYPE II	1		1	EA.
	CURB INLET L-10' TYPE II	1		1	EA.
	BACKFILL & PLUG 18" R.C.P.		45	45	
	REMOVE 18" R.C.P.		19	19	
	PARKWAY CULVERT TYPE B PER O.C.E.M.A. STD. PLAN N° 309	1		1	EA.

APPROVED	DATE

Jack G. Raub Company
Engineering & Planning
P.O. Box 5039   125 Baker Street   Costa Mesa, California 92626
(714) 751-2501

Rudy H. Garcia   R.C.E. 25040   DATE   Mar. 9, 1981

FIELD BOOK		TITLE SHEET		SHEET
DESIGNED	JGS			**1**
CHECKED	WWS	**TENTATIVE**		of 4
SCALE	AS SHOWN	**TRACT 11276**		
DRAWN	JGS			
DATE	3-81	MISSION VIEJO, CALIF.		JN 01-10-035

BASED ON THE
ON THE MAP OF
PAGES 8 THROUGH
OS OF ORANGE
12°00'E

3E-88-71
N THE SOUTHWEST PART
MARGUERITE PKWY 98FT
PLAN OF TRABUCO, 99FT
IA A S.S. BY 6.5.F.T.
WITH THE SIDEWALK

Fig. 7-8

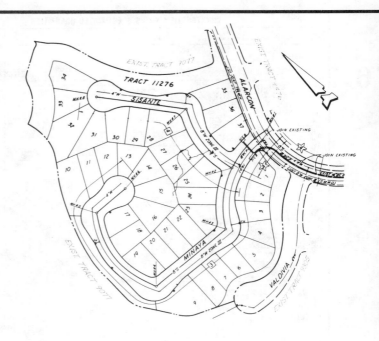

INDEX MAP
SCALE 1"=80'

WATER NOTES CONT.:

12. THE WATER SYSTEM SHOWN HEREON FOR TRACT 11276 IS ENTIRELY WITHIN THE SANTA MARGARITA WATER DISTRICT ZONE III SERVICE AREA.

13. ALL LOTS SHALL HAVE COPPER WATER SERVICES AND BRONZE WATER METERS.

14. ALL ZONE III WATER MAIN SHALL BE CLASS 200 A.C.P.

15. THE PLANS NOT WITHSTANDING, VALVES SHALL NOT BE CONSTRUCTED IN CROSS GUTTER OR GUTTER LOCATIONS.

LEGEND

—○— DENOTES PVC. SEWER MAIN WITH MANHOLE
—6"W— DENOTES A.C.P. WATER MAIN
—⊶ DENOTES BUTTERFLY VALVE
⬍ DENOTES FIRE HYDRANT
—N"RW—○ DENOTES RECLAIMED WATER LINE
—○ DENOTES WATER SERVICE & METER
—■ DENOTES 2"RW. & METER (UNLESS OTHERWISE NOTED)
━━ DENOTES RCP STORM DRAIN
Ⓔ DENOTES CONSTRUCTION NOTES
⊘ DENOTES CURVE DATA
◇ DENOTES STORM DRAIN CURVE DATA
△ DENOTES JUNCTION STRUCTURE NUMBER
③ DENOTES CATCH BASIN NUMBER
❖ DENOTES STREET LIGHT
▲ DENOTES STREET NAME SIGN
▣ DENOTES SHEET NUMBER
★ DENOTES STORM DRAIN SHEET NUMBER
33.71'OC DENOTES DISTANCE ON CURB

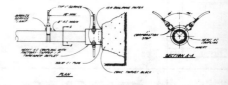

WATER SERVICE CONNECTION FOR DEAD END MAINS
NOT TO SCALE

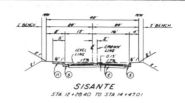

SISANTE
STA. 12+28.40 TO STA 14+47.01

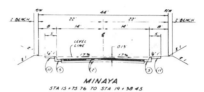

MINAYA
STA 15+75.76 TO STA 19+38.45

MINAYA   **SISANTE**
STA 10+00 TO STA 15+75.76   STA 10+00 TO STA 12+28.40

CONSTRUCTION NOTES:

② CONST AC/AB WITH SC-250 PRIME COAT & SEAL COAT
⑤ CONST CURB & GUTTER TYPE "D" (6" CF) PER OCEMA STD PLAN Nº 201
⑪ CONST SIDEWALK PER OCEMA STD PLAN Nº 205

STREET NAME SIGN LOCATION

LOC	STREET NAME ⊕	STREET NAME
NE	ALARCON	MINAYA
NE	MINAYA	SISANTE

ALARCON
SPECIAL PROVISIONS
OPEN CUTS

Saw cutting of the A.C. shall be made to the satisfaction of the EMA Field Inspector

Two sack cement slurry backfill shall be used in transverse open cuts to within four inches (4") of finish surface and the slurry backfill shall be placed within 24 hours after the trench is opened

All backfill replaced in excavations parallel to the road centerline shall be compacted to 90% relative compaction. Backfill material shall be subject to EMA field inspector's approval prior to placement. EMA field inspector may require sand or 2 sack cement slurry backfill

Patching of the trench shall be of high quality and to the total satisfaction of the Field Inspector. Permanent A.C. shall be placed within two (2) working days after backfilling

Contractor shall be responsible for adjusting finish grade of A.C. patch due to settlement commencing no less than 10 working days, and completing no later than 20 working days after A.C. patch installations as provided above. All work shall be done to the satisfaction of the Field Inspector

Traffic shall be maintained at all times and shall be protected with adequate barricades, lights, signs and warning devices as per the current State of California Department of Transportation Manual of Traffic Controls, and to the directions of the EMA Field Inspector

A minimum of one (1) PAVED LANE SHALL BE provided for traffic in EACH DIRECTION. In addition a minimum clearance of two feet (2') adjacent to any surface obstruction and a five-foot (5') clearance between the excavation and the traveled way shall be maintained

Trench steel plating shall be on the job site before the trenching begins (at the discretion of the Field Inspector)

All plates shall be pinned or tacked to approval of the EMA Field Inspector

All permanent striping shall be replaced in kind

CURB INLET MODIFICATION DETAIL
NOT TO SCALE

ck G. Raub Company				INDEX MAP,
Engineering & Planning		PLAN SCALE AS SHOWN		SECTIONS & NOTES
	DATE	BY	REVISION DESCRIPTION	APP'D	DATE	TRACT 11276

SHEET 2

Fig. 7-9

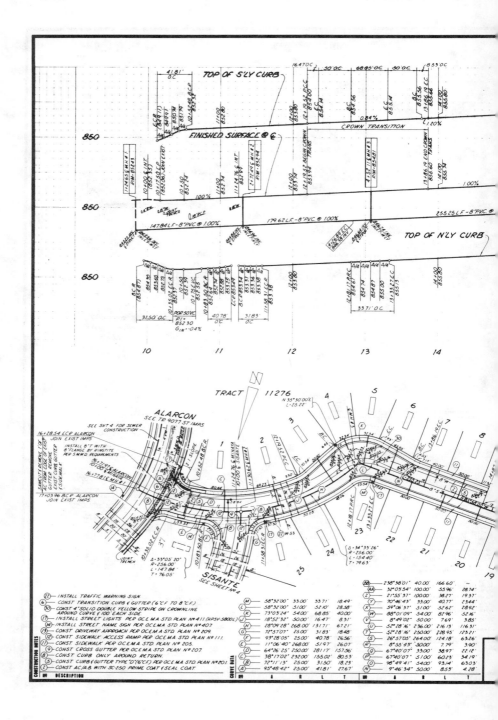

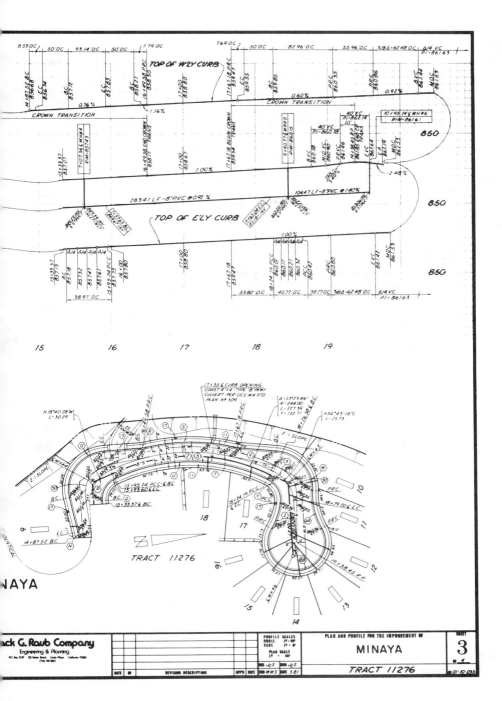

Fig. 7-10

EXERCISE 7-1: Use the field notes in Figure 1-8 and plot the ground level as given and then draw the pipe line as per the C&F notes in the right column. Include showing the manholes as designated. Calculate and show the grade of the pipe line; if there is more than one grade show the break point between grades.

Establish and show the finish grade of the street surface over the line from the elevations for the "RIM" of the manholes:

Sta. 0+00 = 807.51
Sta. 2+55 = 810.89
Sta. 5+20 = 812.95

Elevation of Outlet S. Sta. 5+18 = 807.05.

* * *

PROBLEMS:
1. What is the difference between a "plan" and a "profile"?
2. When are three lines of elevations needed for a profile?
3. When do you need field notes showing *lateral* information on a pipe line job?
4. Name four types of "NOTES" to be found on the title sheet of a set of Plan & Profile drawings.
5. Name two other items that should be on the title sheet.
6. Where are "special items" shown?
7. Name three items sometimes added to the R/W lines shown on the "plan".
8. Why are the scales different for the "plan" and for the "profile"?
9. What special information do you show on the finish grade line between stations?

Chapter 8

Cross-Section Drawings

Cross-section Drawings

As in many cases, a pictorial presentation assists in solving the mathematical problems. For the purpose at hand, the existing ground elevations can be plotted directly from cross-section field notes or they can be picked off the contour map. After the designer has established the finish grade, you take the rough grade and finish grade lines plus the outside points where the cut or fill will meet the natural ground and draw cross-section graphs to scale so that those areas can be used to calculate the total amount of earth to be moved.

The picture of the cross-section drawing is made normal to the direction of the center line or control line. In Figure 8-1, two typical sections are shown of the relationship of the existing ground to the finish grade, one in cut and the other with fill.

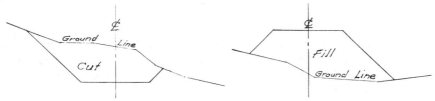

Fig. 8-1

The same principle of showing cuts and fills applies whether it is for a reservoir, railroad, highway, canal or flood control channel.

The slope, either up or down, from the finish grade to the adjacent existing ground is defined in a mathematical relationship. The expression of 2½ to 1, or shown as 2½:1, means that for each 2½ feet (or meters if in metric scale) of horizontal distance, the vertical difference up or down equals 1 foot (or meter). Obversely, a 1:2 slope means that for each 1 foot (or meter) of horizontal distance, the vertical difference is 2 feet (or meters). Of course, the latter would be much steeper than the former.

From Cross-section Field Notes.

Using Figure 1-9, the notes taken at the cross-section station 0 + 50 have been reduced and the resulting ground line is shown here in Figure 8-2 in conjunction with a cut designed for a flood control channel.

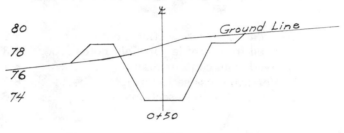

Fig. 8-2

EXERCISE 8-1: From the information given at station 2 + 25 of the field notes in Figure 8-3, reduce the rod readings to elevations and plot them on cross-section paper to show the existing ground line using a horizontal scale of 1″ = 10′ and a vertical scale of 1″ = 5′. A vertical scale different from that of the horizontal is used to accent the elevation differentials. Label the lines representing the elevations of 115, 120 and 125.

+	H.I.	−	Sta		West	N+S ℄	East	
	126.43		2+25		9.8 0.8 7.4 6.8 6.0 50 41 36 20 7	5.2 6.1 6.8 7.4 10.3 11.1 12 20 33 40 50	
			2+00				

Fig. 8-3

Draw a *Cut* Grade Elevation line at 116.2 for 10 feet each side of the center line. At each end of the *Cut* Grade line 10 feet from center line, establish a 3:1 slope line up to the existing ground line and label it.

* * *

A series of drawings of this type of representation is used for calculating the amount of earth to be cut or filled in the project.

Cross-sections are usually plotted on 10 units to the inch cross-section paper and arranged consecutively by the stations which were reported in the field notes. The first station of the project may be placed in either the lower left or upper left corner of the sheet and the others follow up or down the sheet. Because this type of work seldom leaves the office, pencil is usually satisfactory.

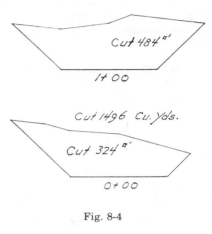

The station of the cross-section is shown below it and the resultant square footage is marked within it. The volume between two cross-sections — usually in cubic yards — is shown on the sheet between the consecutive cross-sections all as illustrated in Figure 8-4.

Fig. 8-4

The procedure for calculations will be discussed in Chapter 9.

From Profile-run Field Notes.

Whereas the cross-section type of field notes are usually based upon a preset grid pattern, the profile-run type of field notes are, as the words imply, notes reporting the elevations usually on a single line. These alone are not sufficient for drawing cross-sections but in some cases the profile notes carry offset information on each side of the profile line as shown in Figure 1-19.

EXERCISE 8-2: Reduce the rod readings for all points read at Station 54 + 50 in the notes shown in Figure 1-19 and plot them on cross-section paper at a horizontal scale of $1'' = 25'$ and a vertical scale of $1'' = 2'$.

Set the *Cut* Grade Elevation line at 248.0 feet for 8 feet each side of center line. At each end of the *Cut* Grade line, establish a 1:1 slope line up to elevation 254.0 and label it, then join the left side with the ground elevation of 254.0 and the right side with the ground elevation of 253.4.

* * *

From Contour Maps.

There are times when preliminary studies are requested, before any field survey work is done, concerning the amount of earth moving that would be required for one route compared with another route.

Again the engineer resorts to the use of quadrangle maps for his comparative study. An alignment is drawn on the map and cross-sections are taken at critical points. In this case, fewer cross-sections are used than in the final analysis.

An example of one segment for such a study is shown here. Figure 8-5 is the plan-profile delineation and Figure 8-6 shows the three cross-sections taken from it.

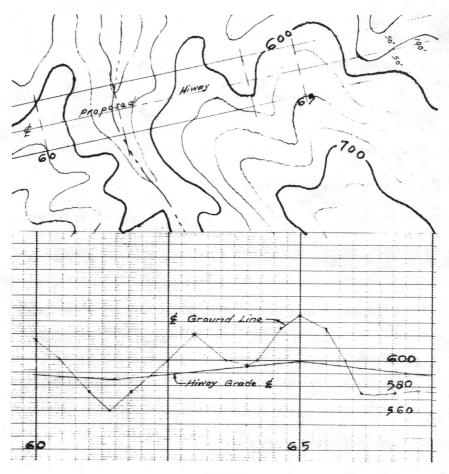

Fig. 8-5

Vertical Cross-sections

You have seen that some cross-section drawings are made with diverse scales to accent the vertical differentials and others are drawn with the same horizontal and vertical scales.

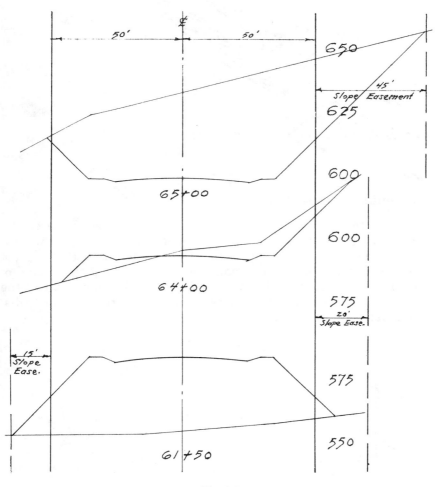

Fig. 8-6

When the area of the cross-section is calculated from the elevations and distances given, or by geometric deductions, the scale does not matter as long as you are using the figures. If, however, you use a planimeter on the outline of the cross-section to determine its area, then the scale must be the same for both horizontal and vertical directions.

A perspective drawing of the three cross-sections of cut and fill with transition lines is shown in Figure 8-7.

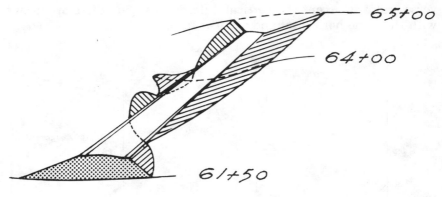

Fig. 8-7

Horizontal Cross-sections

Still another method is used for determining cross-section areas for the purpose of calculating earth volumes; it is making use of the horizontal area of successive planes of elevations. By following the circuit of a single contour line with a planimeter, or breaking that plane of elevation into parts for additive calculations, you can determine the area within it, then it serves the same purpose as the previously discussed vertical cross-sections.

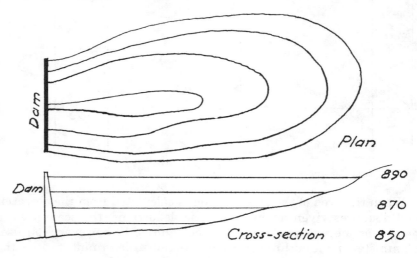

Fig. 8-8

In the case of creating a reservoir by building a dam across a river, it is desired to know the volume of water which the dam will impound. Look at Figure 8-8 and see how following each circuit of selected contour lines upstream from the dam, you will have a series of vertical cross-section areas from which the volume can be derived.

This same application can also be used for calculating a dirt fill in a sump or for cuts-and-fills of building sites.

PROBLEMS:

1. In the fraction or proportion of 2:5, what does the first number represent? — what does the second number represent?

2. Why should the vertical scale on a profile differ from the horizontal scale?

3. Do you use different scales or the same scale for horizontal and vertical distances on cross-section drawings?

4. Do you ever draw and compute cross-sections without the use of field notes?

5. In lieu of drawing cross-sections, how would you figure the volume of water in a reservoir?

— NOTES —

Chapter 9

Earthwork Calculations

Earthwork Calculations.

The volume of earth to be moved, concrete to be placed in construction, quantities in stockpiles, capacity of tanks or a reservoir of water impounded by a dam, all are obtained by this type of calculation. There are several methods of calculating such volumes and they are applicable in horizontal and/or vertical directions.

The areas of the cross sections described in Chapter 8 can be computed through arithmetic, geometry or by planimeter. The volume between sections, or through several sections, can be developed by either the longitudinal or the prismoidal formula. The obverse method of using horizontal areas is by planimetering successive contours within the confines of the project and extrapolating the volume from them through either formula.

The unit of volume is either a cubic foot or cubic yard of which the latter is the standard unit for concrete and earthwork.

The volumetric shape between the cross-sections approaches the forms of prisms but because they are rarely, if ever, of a regular prismatic form, they are called prismoids.

Longitudinal Theory.

This is also known as the average end area method and it is the most commonly used because it is easier to apply. The computed volume is usually greater than actual and therefore favors the contractor.

The principle is based on the formula that where the ends are parallel with each other, the volume of a solid in cubic feet equals the average of the areas of each end in square feet times the distance between them expressed as: $V = \left(\dfrac{A_1 + A_2}{2}\right) D$

taken from Figure 9-1.

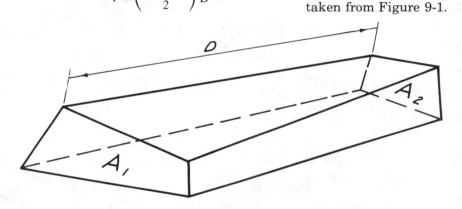

Fig. 9-1.

The volume in cubic yards would be: $V = \dfrac{D}{54}\left(A_1 + A_2\right)$

For the study of calculating the end area cross-sections, they can be classified as single-level, three-level, five-level and irregular. The single-level or regular section is where the field levels have been taken at one point on each side of the centerline and this permits numerical calculations to be made without plotting as shown in Figure 9-2. It is equally applicable to cut or fill.

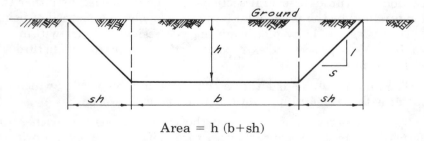

Area = h (b+sh)

Fig. 9-2

If the ground is not parallel with the finish grade, then you have one of the conditions called a three-level situation as shown in Figure 9-3 and it is necessary to use a combination of triangle formulas. Again it is good for either cut or fill using the four triangles shown above (or below) the base line.

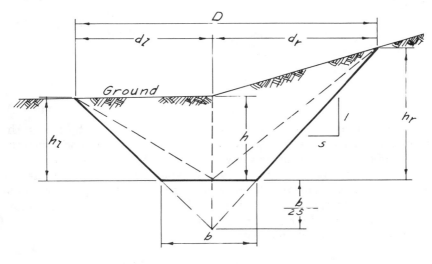

Fig. 9-3

An alternate method which is shorter is to calculate a total area formed by prolonging the slope lines to their intersect and then subtract the triangle thus formed which is outside the roadbed. Referring again to Figure 9-3:

$$\text{Area} = \frac{1}{2} \left(h + \frac{b}{2s} \right) \times (d_1 + d_r) \; - \; \frac{1}{2} \left(b \times \frac{b}{2s} \right)$$

$$= \frac{1}{2} \left(h + \frac{b}{2s} \right) \times D - \frac{b^2}{4s}$$

Although this would apply to any combination of slope and width of base, if these two elements are constant for any segment of the construction, you need to calculate the "$\frac{b^2}{4s}$" only once and subtract the same figure from each overall calculation.

The next form of complex cross-section is called the five-level situation shown in Figure 9-4 using a combination of triangles having common vertical sides. Combining the formulas for each pair of triangles, you have:

$$\text{Area} = 1/2 \; (h_1 d_1 + hb + h_r d_r)$$

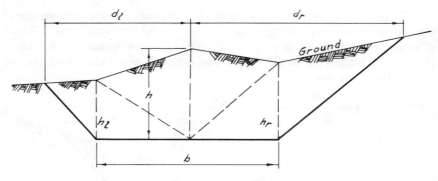

Fig. 9-4

The irregular cross-section is the situation where the breaks in the ground elevations are many and thus seldom, if ever, coincide with the breaks of the finish grade as shown in Figure 9-5. Three of the ways to determine the area of such a polygonic shape are: 1 — compute the trapezoidal areas and from their sum subtract the triangular areas at each end, 2 — compute the areas by coordinates, 3 — divide the area into triangles and add their algebraic sum.

By the first method you have:

$$A = \left(\frac{BB' + CC'}{2}\right) B'\,C' + \left(\frac{CC' + DD'}{2}\right) C'D' + \left(\frac{DD' + EE'}{2}\right) DE' +$$

$$\left(\frac{EE' + FF'}{2}\right) E'\,F' + \left(\frac{FF' + GG'}{2}\right) F'\,G' - \left[\left(\frac{BB'}{2}\right) B'\,A + \left(\frac{GG'}{2}\right) HG'\right]$$

$$= \left(\frac{19}{2}\right) 11 + \left(\frac{23}{2}\right) 13 + \left(\frac{34}{2}\right) 14 + \left(\frac{37}{2}\right) 10 + \left(\frac{31}{2}\right) 5 - \left[\left(\frac{9}{2}\right) 9 + \left(\frac{15}{2}\right) 14\right]$$

$$= 104.50 + 149.50 + 238.00 + 185.00 + 77.50 - [40.50 + 105.00]$$

$$= 754.50 - 145.50 = 609.00$$

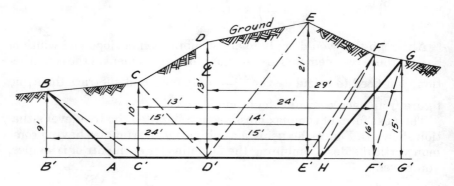

Fig. 9-5

By the second method using the field notes of this particular cross-section:

¢

A	B	C	D	E	F	G	H
0	C9	C10	C13	C21	C16	C15	0
15	24	13	0	14	24	29	15

You have coordinates based on the finish grade expressed by x coordinates measured horizontally from the centerline axis and y coordinates measured vertically from the finish grade axis. With the 8-sided figure shown, you can use the formula:

$$\text{Area} = \tfrac{1}{2} \, [Y_1 (X_2 - X_8) + Y_2 (X_3 - X_1) + Y_3 (X_4 - X_2) + \ldots \ldots]$$

For any figure, you have:

$$\text{Area} = \tfrac{1}{2}[x_1 y_2 + x_2 y_3 + \ldots x_1 y_{(i+1)}] - [x_2 y_1 - x_3 y_2 - \ldots x_{(i+1)} y_1]$$

Which tabulates as follows: $y_n (x_{n+1} - x_{n-1}) = \quad 2 \times \text{area}$

y_n	$(x_{n+1} - x_{n-1})$			$2 \times$ area
0	$(-24 - 15)$	=		0
9	$(-13 + 15)$	=	+	18
10	$(\ 0 + 24)$	=	+	240
13	$(\ 14 + 13)$	=	+	351
21	$(\ 24 + \ 0)$	=	+	504
16	$(\ 29 - 14)$	=	+	240
15	$(\ 15 - 24)$	=	−	135
0	$(-15 - 29)$	=		0

$$1218 \div 2 = 609$$

By dividing the area into triangles, you can cross multiply the horizontal and vertical distances and divide by 2. Of course, if any extend outside the cross-section proper, you subtract the excess. Again referring to Figure 9-5, the calculations are thus:

9	(2)	=	18
10	(24)	=	240
13	(27)	=	351
21	(24)	=	504
16	(14)	=	224
16	(1)	=	16
			1353
−15	(9)	=	135

$$1218 \div 2 = 609$$

Referring back to the field notes at the top of page 9.5, you can tabulate them by the coordinate system under x and y and cross multiply as follows.

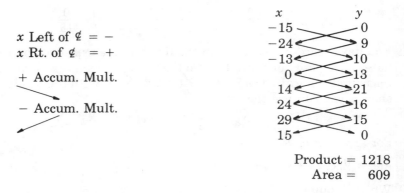

x Left of ₵ = −
x Rt. of ₵ = +

+ Accum. Mult.

− Accum. Mult.

x	y
−15	0
−24	9
−13	10
0	13
14	21
24	16
29	15
15	0

Product = 1218
Area =   609

Where you have transitions from cut to fill or vice versa as shown in Figure 9-6, calculate each part separately for cut and for fill rather than try to combine them.

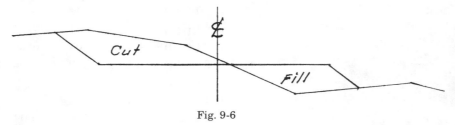

Fig. 9-6

Now that you have the method of determining the areas of various configurations of cross-sections, you can apply them to the average end area formula shown with Figure 9-1. This process is repeated for every segment along the route in order to obtain the total volume involved in the project.

Borrow Pit Theory.

This method is mainly used for determining the volume of limited areas, in contrast with a continuous route. It is applied to such items as stockpiles of coal, gravel, etc., excavation of a basement or a large area for a building, leveling for a tennis court, etc., all of which have one flat elevation and a series of variable alternate elevations. The volumes can be compared to truncated rectangular and/or triangular prisms.

A simple illustration of just a rectangular area for a basement excavation is shown in Figure 9-7.

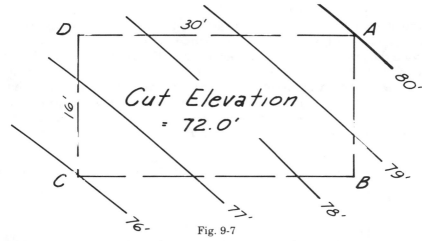

Fig. 9-7

In perspective, the cut would appear like the truncated rectangular prism shown in Figure 9-8. An approximation of the volume could be obtained by multiplying the cut elevation area by the average depth of the excavation. By interpolation and differentials, the amounts of cuts would be A = 8.0', B = 6.6', C = 4.0' and D = 5.6'; therefore, the average cut would be: $\dfrac{8.0' + 6.6' + 4.0' + 5.6'}{4} = 6.05'$

The area of 16' × 30' = 480 square feet times the average cut of 6.05 feet equals 2904.0 cubic feet or 107.6 cubic yards.

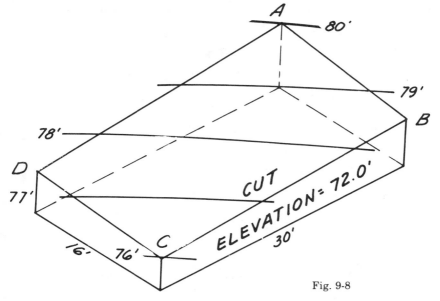

Fig. 9-8

In a large area with greater differences of elevations, the above method would not give as accurate results as the following. For a comparative analysis we will use the same facts as shown in Figure 9-7. Divide the area into rectangles and assign letters to the corners of each rectangle as follows: 'a' to each corner that occurs in only one rectangle; 'b' to each corner that occurs in two rectangles; 'c' to each corner that occurs in three rectangles; 'd' to each corner that occurs in four rectangles. Interpolate between the contours and assign such elevations to the corners of each rectangle as shown in Figure 9-9.

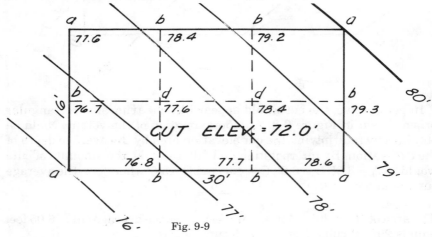

Fig. 9-9

Tabulate the sums of the depth of cut at each corner as follows:

Sum of 'a': $\dfrac{76}{72}$ $\dfrac{77.6}{72}$ $\dfrac{80}{72}$ $\dfrac{78.6}{72}$
$4.0 + 5.6 + 8.0 + 6.6 \qquad\qquad = 24.2$

Sum of 'b': $\dfrac{76.7}{72}$ $\dfrac{78.4}{72}$ $\dfrac{79.2}{72}$ $\dfrac{79.3}{72}$ $\dfrac{77.7}{72}$ $\dfrac{76.8}{72}$
$4.7 + 6.4 + 7.2 + 7.3 + 5.7 + 4.8 = 36.1$

Sun of 'c': $\qquad\qquad\qquad\qquad\qquad\qquad\qquad = 0$

Sum of 'd': $\dfrac{77.6}{72}$ $\dfrac{78.4}{72}$
$5.6 + 6.4 \qquad\qquad\qquad\qquad = 12.0$

The formula for volume in this case will be:

$$V = \text{area of one square} \times \frac{\text{'a'} + 2\text{'b'} + 3\text{'c'} + 4\text{'d'}}{4}$$

$$= 8 \times 10 \times \frac{24.2 + 2(36.1) + 3(0) + 4(12)}{4}$$

$$= 80 \times \frac{144.4}{4} = 2888 \text{ cu. ft.} = 107.0 \text{ cu. yds.}$$

Although the difference here is only about 1%, it is obvious that in a matter of a 7500 yard job, the cost of 75 yards could be worth the more accurate calculation.

Since the formula for volume in cubic yards is:

$$V = \frac{A}{4 \times 27} \ ['a' + 2'b' + 3'c' + 4'd'] \ \ldots \ldots$$

if the ground grid is laid out in rectangles of 27' x 40' or 30' x 36', then

$$\frac{A}{4 \times 27} = \frac{1080}{108} = 10$$

so if the elevations are taken to the nearest tenth of a foot, then the sum of the multiplied corner heights with the decimal point removed is the volume direct in cubic yards.

All excavations do not have just rectangular patterns. Let us consider the situation shown in Figure 9-10.

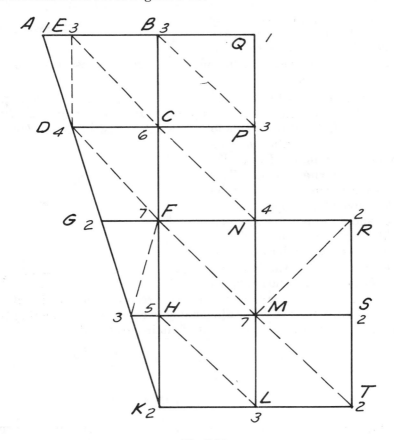

Fig. 9-10

Where large areas are involved with considerable differences in ground elevations, each marked area should be calculated as a truncated prism and the sum of their volumes added for the total.

If the surface area cross-sectioning results in a trapezoid, because of a diagonal, such as with ABCD in Figure 9-10, it should be broken into the triangle AED and rectangle EBCD for the purpose of calculating the volumes of the truncated triangular prism AED-A'E'D' and the truncated rectangular prism EBCD-E'B'C'D as illustrated in Figure 9-11

With rolling ground, one plane cannot usually be passed through the four-corner elevations but one plane can be passed through three points, so by dividing a four-point area (square, rectangular or trapezoidal) with a diagonal in either direction, two triangular accommodating planes are created. The surveyor in the field

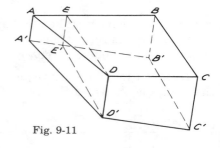

Fig. 9-11

may indicate the appropriate diagonal to use from his observation of the prevailing conditions.

Referring again to Figure 9-10, the dash lines show the field grid areas broken into triangles. The numbers at each corner show how many times each cut (or fill) is to be used in the calculation; this represents the number of triangles that meet at that corner. The numbering of the corners can be checked by adding them all together and dividing by six which should equal the number of rectangles under consideration. To develop the overall formula which will incorporate all of the triangles

C_1 C_2

C_4 Fig. 9-12 C_3

in such a situation, let us start with a single rectangle such as $C_1 C_2 C_3 C_4$ shown in Figure 9-12 with the diagonal between C_2 and C_4 making two triangles, then the volume in terms of cubic yards will be:

$$V = \frac{A}{2 \times 27} \left(\frac{C_1 + C_2 + C_4}{3} + \frac{C_2 + C_3 + C_4}{3} \right)$$

$$= \frac{A}{6 \times 27} \; (C_1 + 2C_2 + C_3 + 2C_4)$$

Therefore, the volume in cubic yards for a group of rectangular prisms with their divisions into triangular prisms would be:

$$V = \frac{A}{6 \times 27} (\Sigma c_1 + 2\Sigma c_2 + 3\Sigma c_3 + 4\Sigma c_4 + 5\Sigma c_5 + 6\Sigma c_6 + 7\Sigma c_7 \ldots \ldots)$$

The subscripts represent the number of triangles that meet at any one corner. While this would apply to any group of rectangles, in the case of the configuration shown in Figure 9-10, the triangular and/or trapezoidal prisms would need to be calculated separately and added to the group calculation.

If the ground grid is laid out in rectangles of 36' x 45' or 40' x 40.5', then $\frac{A}{6 \times 27} = \frac{1620}{162} = 10$

so if the elevations are taken to the nearest tenth, the sum of the multiplied cuts (or fills) with the decimal point removed is the volume direct in cubic yards.

Prismoidal Theory.

Volume obtained by this method is more exact because the formula includes consideration of a middle cross-section in addition to the end areas. This mid-section is either taken in the field or developed from the data on the end sections. When it is not given in the field notes but extrapolated in the office, it is NOT the *average area* of the two end areas; it is an area developed from factors which are equal to the means of corresponding dimensions at the two adjoining end sections.

The prismoid is a solid with parallel end areas that may be composed of any combined make-up of other solids such as prisms, pyramids or wedges of which their bases or apexes lie in the end areas. A simple trapezoidal prism was shown in Figure 9-1. A little more complex form, the three-level end section prismoid, is shown in Figure 9-13 with end areas divided into triangles

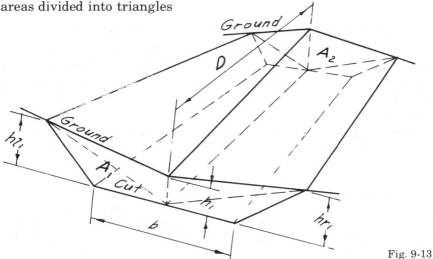

Fig. 9-13

The prismoidal formula for volume in cubic feet is:

$$V = \frac{D}{6} (A_1 + 4A_m + A_2)$$

where A_m is the cross-section area of the mid-section.

Except for ground irregularities not reflected in the field notes of the cross-section, this formula gives the more nearly correct volume.

Because the average-end-area method is faster than the prismoidal application, it is often used with a prismoidal correction. The correction in cubic feet for a triangular-end-section solid as in Figure 9-14 is:

$$C = \frac{D}{12} (b_1 - b_2)(h_1 - h_2)$$

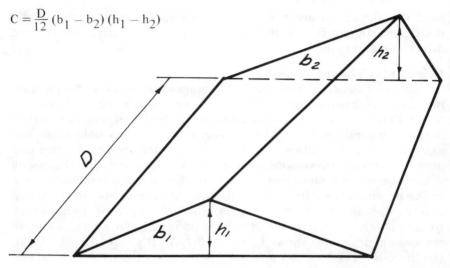

Fig. 9-14

This is to be subtracted from the average-end-area volume.

When the more complex forms are broken into triangular prisms, this correction can be applied to each one. In the configuration shown in Figure 9-13, or any similar one, the corrections on the lower triangles are zero and the obverse is true in the case of *fill* where the upper triangles have constant factors.

Some handbooks for railroad or highway engineering contain tables for prismoidal corrections as well as tables for earthwork volumes and these offer some saving in time.

Area by Planimeter.

A method of deducing irregular areas by mechanical means has been developed in an instrument known as a planimeter, one type of which is shown in Figure 5-18. From an accurately scaled drawing, you carefully follow the outline by a tracer (T), that is either a metal pointer or a dot on

a magnifying glass lens, attached to the arm (AT) ex-
tending from the "counter
box" (c) illustrated in Figure
9-15. This box has a socket
into which the pivot (B) at
the end of the polar bar (PB)
rests and allows the two arms
to rotate around the pole (P)
which is a fixed point. Under
the "counter box" is a wheel
in contact with the paper
which revolves forward or
backward as the pointer follows the outline. The number
of net revolutions is registered by a dial to the nearest

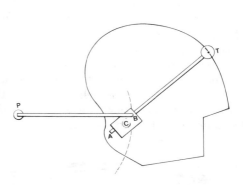

Fig. 9-15

0.01 revolution and a vernier dial registers the nearest 0.001 revolution.

Before setting the pole (P), try moving the tracer (T) around the entire
perimeter of the area in question to be sure it can be traced completely
from one setting. If the area is too large for one swing around it,
establish one or more division lines, planimeter each part and add them
together.

There is also available now a larger model of planimeter with approximately 4 inch diameter wheels and extended arms to cover larger sized
maps and which is tied electronically to a computer that gives a direct
read-out of the area traversed.

For translating the revolutions of the counter wheel into area, it is
necessary to determine and establish the planimeter's "constant" which
means the ratio between the revolutions of its wheel and an outlined
area. If the length of the tracer arm (AT) is set without the ability to
adjust, it is usually set at a ratio of 1:10 which means that tracing
around a 1" x 1" square, the dial will record a 1.00 revolution; therefore,
the area traced on the paper equals the number of revolutions multiplied by 10; this then is a constant of 10.

Most planimeters have an adjustable arm, usually with a scale
marked on it, which permits different settings such as 1" = 20' up to 1" =
200'. Set for a linear scale of 1" = 20' means an area of 1 square inch =
400 square feet.

To take a reading on the area to be planimetered, you may either set
the initial dial reading at zero, or, take the dial reading as it is at the
beginning point of the tracing. By the first method, your final reading
will be the one for calculating the area; by the second method, you will

need to take the difference between the first and last readings for your area calculation.

For greater accuracy, it is advisable to trace the outline twice, and preferably once in reverse direction, to get two readings and average them.

Although not so often used, the previously discussed end areas of cross-sections are sometimes determined by the planimeter method. Since the principle of the planimeter is to measure square inches (or square centimeters) which reading is then converted to the required units by multiplication, the horizontal and vertical scales may be either the same or different as when used to exaggerate the slope.

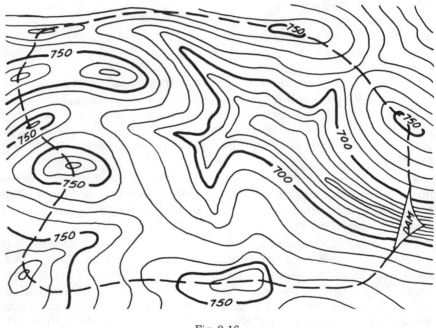

Fig. 9-16

Another application suitable to planimeter work is the determination of areas enclosed by contour lines. Figure 9-16 shows a mountainous area with a proposed dam for a reservoir and the drainage basin outlined by a dash line. It is desired to know the drainage area and the reservoir capacity behind the dam that will be built to hold water up to the 700-foot level. This can be computed by the same formula used for the average-end-area volume; in other words, since the contour areas are parallel with each other (vertically), the volume or capacity in *cubic feet* between contours equals the average of the areas of successive contour

$$V = \left(\frac{C_1 + C_2}{2}\right) d$$

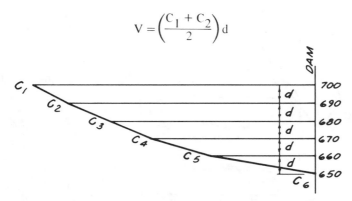

CROSS-SECTION OF CONTOURS
AT DAM

Fig. 9-17

etc. for each segment or combining them into one equation you have:

$$V = \left(\frac{C_1}{2} + \Sigma C + \frac{C_n}{2}\right)^{n-1} \quad \text{or} \quad V = \left(\frac{C_1 + C_6}{2} + C_2 + C_3 + C_4 + C_5\right)$$

so the total capacity is the sum of the volumetric series of levels in cubic feet.

Differential Elevation Calculations.

At ten-foot intervals, however, this computation would be only an approximation. For the final determination, the volume should be computed on the basis of one or two-foot intervals (contours) and the prismoidal formula used for the number of *even* layers (lying between an odd number of contours) would be in cubic yards based on d in feet and c in square feet:

$$V = \frac{d}{3 \times 27} (C_1 + 4C_2 + 2C_3 + 4C_4 + 2C_5 + \ldots C_n)^*$$

and C_n *must* be the odd contour. If there is an odd number of layers (which has an even number of contours), one of them must be calculated separately and added to the other total.

Another application of this principle is shown in Figure 9-18 where a building site is cut into a hillside. In this case both the natural contour lines and the finish contour lines are drawn and the points where lines of

*Based on the prismoidal formula and Simpson's one-third rule per Johnson's THEORY AND PRACTICE OF SURVEYING.

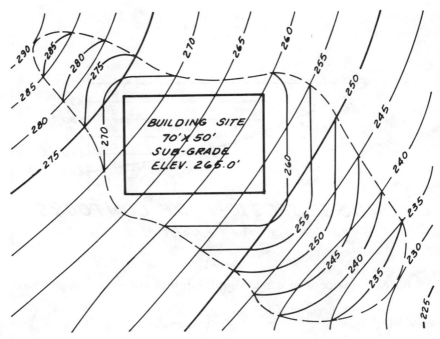

Fig. 9-18

elevation of the
natural grade and
finish grade inter-
sect are joined by a
dash line. These
points are terminals
of the cut or fill con-
dition. The exception
to this is where the
elevation of the
building site coin-
cides with the
natural grade and
the dash line is held
away from the site
enough to give a
shoulder to it. Figure
9-19 shows a "slice"

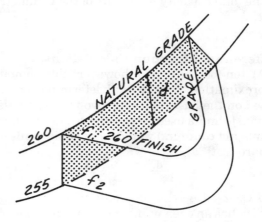

A "slice" of fill in perspective.
Fig. 9-19

of the fill area between elevations 255 and 260. The prismoidal formula
last cited will again apply to any *even* number of "slices." Also, more
accuracy will be obtained by the use of one or two-foot contours. C_1, etc.

in the formula would be for the upper area in cut and F_1, etc. in the formula would be for the lower area in fill.

EXERCISE 9-1: Using the above formulas, compute the volume of cut and volume of fill from the contours shown on the drawing in Figure 9-18. Run the planimeter on the contour lines with the same elevation, for cut and fill, as are attached to the natural contour line. Use cross-section method to check your calculations by planimeter.

* * *

Now let us apply this principle to a roadway which must be cut through one part of a mountain and over fill in a canyon adjoining as shown in Figure 9-20.

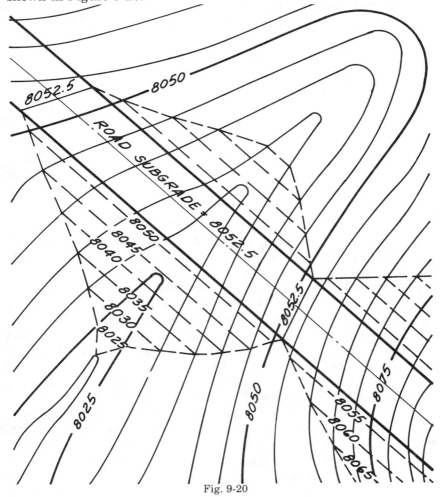

Fig. 9-20

Again the intersection points of the natural contours with the finish contours are connected by a dash line to indicate the outlines of the fill areas and the cut areas.

The volume can be computed from either vertical cross-sections at right angles to the roadway or horizontal layers between contours, the latter of which would be more easily done by planimeter.

Other Forms of Areas and Volumes.

Besides the applications discussed so far, there are other situations which require a comprehension of the shape and the formula applicable to it for determining the area and/or volume required. Let us consider the irregular part of an overall area ACDEH which has a water boundary as illustrated in Figure 9-21.

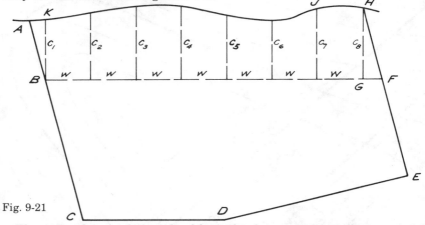

Fig. 9-21

There are three rules applicable to the determination of an area such as KBGH within the total irregular area of ABFH. They are known as 1 - the *Trapezoidal Rule,* 2-*Durand's Rule* and 3- *Simpson's Rule*, sometimes called the *One-Third Rule*. Applying the same information to each one, it can be shown that the Trapezoidal Rule is the least accurate, Simpson's Rule is the most accurate and Durand's Rule is in between.

Another factor in determining the amount of accuracy is the number of offsets used. It is obvious in Figure 9-22 that neither the trapezoid NOPT nor NOQS fits the curve PRT very well. Also, the trapezoid LMNT does not fit the curve LUT very well either, but if two trapezoids are formed as LMVU and UVNT they approximate the curve LUT much better. To accommodate all the curves on the ground by additional field work of measuring more offsets is often more expensive than it is worth. If the offsets are taken at irregular spacing, you have no alternative but to calculate each area, triangular or trapezoidal, and add them for the total. If, however, the spacings between offsets are equal, you may need to compute only the end triangles (if any) separately and the remainder area by one of the following three formulas using the designation shown in Figure 9-21.

The *Trapezoidal Rule* is:

$$\text{Area} = W\left(\frac{C_1}{2} + \sum_{2}^{n-1} C + \frac{C_n}{2}\right)$$

Where: W = common distances between offsets.

C_1 and C_n = first and last offsets in the group of trapezoids.

$\sum_{2}^{n-1} C$ = sum of the intermediate offsets.

applied: $\text{Area} = W\left[\left(\frac{C_1 + C_8}{2}\right) + C_2 + C_3 + C_4 + C_5 + C_6 + C_7\right]$

Durand's Rule is:

$$\text{Area} = W\left[0.4\,(C_1 + C_n) + 1.1\,(C_2 + C_{n-1}) + \sum_{3}^{n-3} C + C_{n-2}\right]$$

Where: W = common distances between offsets.

C_1 and C_n = first and last offsets in the group of trapezoids.

C_2 and C_{n-1} = second and next to last offset in the group.

C_{n-2} = second from last offset in the group.

applied: $\text{Area} = W\left[0.4\,(C_1 + C_8) + 1.1\,(C_2 + C_7) + C_3 + C_4 + C_5 + C_6\right]$

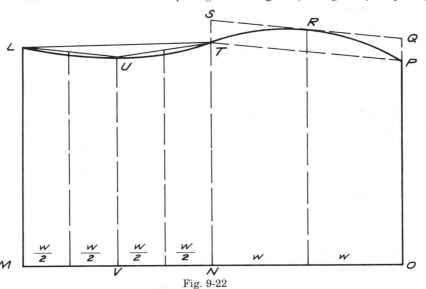

Fig. 9-22

Simpson's Rule is based on the theory that the curved boundary is an arc of a parabola and the total area is composed of the trapezoid NOPT (Fig. 9-22) plus the segmental arc area PRT. This applies only to an even number of offsets which means an odd number of sections. (If there is an even number of sections, one will need to be calculated separately and added.)

$$\text{Area} = \frac{W}{3}\left(C_1 + 2\Sigma C_{odd} + 4\Sigma C_{even} + C_n\right)$$

Where: W = common distance between offsets.

C_1 and C_n = first and last offsets in the group of trapezoids.

$2\ \Sigma\ C_{odd}$ = twice the sum of all the odd numbered offsets.

$4\ \Sigma\ C_{even}$ = four times the sum of all the even numbered offsets.

applied: Area $= \frac{W}{3}\ [C_1 + 2\ (C_3 + C_5 + C_7) + 4\ (C_2 + C_4 + C_6) + C_8]$

Let us turn now to solids. Not only will you find odd-ball forms of which you will need to derive volumes, even though you may have to break it into several different shaped solids, but sometimes the surface areas of them will be required as well; therefore, the following will prepare you for those calculations.

EXERCISE 9-2: Draw a picture in perspective of the following named solids and find and record the appropriate formula opposite each designation. By your doing the drawing, it will help your mind tie the formula to the picture.

Cube:
    Surface —
    Volume —

Rectangular Parallelopiped:
    Surface —
    Volume —

Prism or Cylinder:
    Surface —
    Volume —

Pyramid or Cone:
    Surface —
    Volume —

Frustrum of Pyramid or Cone:
    Surface —
    Volume —

Prismatoid:
    Volume —

Spheres:
    Total area —
    Total Volume —
    Volume of a Segment with one base —
    Zone area —

*   *   *

QUESTIONS:
1. What is the other name for the Longitudinal Theory?
2. Name four classifications of cross-section configurations.
3. Describe the "shorter" method of calculating a cross-section area.
4. Name three different ways of determining the area of an irregular cross-section.
5. What is the Borrow Pit Theory?
6. What is the Prismoidal Theory?
7. Give two reasons why the end area method is sometimes used instead of the Prismoidal Theory.
8. If you have an irregular boundary (along a river for example) and you have equidistant offsets, name the three rules by which you can calculate the irregular area in lieu of calculating each segment separately.

— NOTES —

Chapter 10

Developing a Grading Plan for a Building Site

Grading Plans.

While the plans and profiles previously discussed are used for improvements within strips of land, in contrast with them, the development of building projects requires grading plans covering a limited polygonic area. They do, however, likewise require cross section examination of the relationship of the natural ground compared to the finish grade to be established and calculations made therefrom to determine the amount of earth to be moved.

Although some projects may be concerned with only a single area, others, such as shopping centers or university campuses, may be a group of areas forming a complex at different levels and with connecting driveways. In such a case the addition of vertical curves will be required at the transitions.

Application of Field Notes.

It is conceivable that the "TRANSIT STADIA TOPO" notes shown in Figure 1-26 were taken for the development of that area, rearrangement of the buildings, grading the ground for better use as well as realignment of the road and slopes for drainage.

For the map which you developed from the annotated notes in Figure 1-32, you also followed the stadia form of taking topography.

The exercise you did in Chapter 6 with the field notes for a hillside lot in Figures 6-2, 6-3, 6-6 and 6-7 was a little different application by the use of four control lines plus property lines with offsets for various items.

In every case, the resultant map is a picture of the contour elevations and any pertinent topography.

In the case of a city lot, you may have two sets of notes for its development; one will be the boundary determination as shown in Figure 1-5 and the other will cover the elevations on a grid system as illustrated in Figure 1-10. Whether the topo or elevations are taken by stadia or on a grid depends on the terrain, and therefore the time involved, and the decision is usually made by the field chief. Both of these sets of field notes have to do with Lot 47 in Tract No. 5. According to the recorded dates, all the work was done on the same day; however, either part can be done separately and at different times, although the boundary work is preferably done first in case there is any problem of walls or other encroachments across the property lines. From the grid of elevations shown in Figure 10-1, you would interpolate for the contours and draw them on a map to the scale requested by the architect or engineer.

Looking at Cut and/or Fill.

Let us proceed through an example project and observe the step-by-step development of it. At this point we will assume the field notes have been reduced and plotted and the contours have been extrapolated and drawn as shown in Figure 10-1. A 100 foot × 46 foot × 5 ½-inch concrete slab is to be laid in the rectangle set at 55 feet south of the north line of the property and 50 feet west of the east line of the property.

The cut and fill must be such that no more than 20% of the slab rests on fill dirt. There is to be a shoulder, 4 feet wide on the north and east sides and 10 feet wide on the west and south sides of the concrete slab graded to the same elevation as the subgrade under the slab. Any fill under the area encompassed by the subgrade elevation is to be compacted according to County Specifications No. 328. A driveway is to be laid out between the south shoulder and Millcreek Road with no cut or fill and with a slope of no more than a 5% grade. All cut area is to be at a slope of 4:1 and all fill area is to be at a slope of 8:1.

Excess dirt may be added to the northwesterly and westerly portion of the property.

Since the amount of fill in this project cannot cover more than 20% of the slab area, the cut-off elevation — that is, the level for the subgrade

— must be determined first. Looking at Figure 10-1, the east line of the rectangular 20% area on the west end crosses the 942 foot contour and almost reaches the 943 line at the north edge of the slab. If that east line is turned counterclockwise so that the north end is seven feet west of and the south end is seven feet east of the termini of its first position, the result will be a trapezoidal form for the 20% fill area with the diagonal line approximately parallel with the 942 foot contour across the slab area. If you were to establish a more exact relationship, an interpolation shows that the diagonal line would approximate the 942.3 foot level; however, the even 942 contour is close enough and easier to work with as well as being on the safe side of the fill limit.

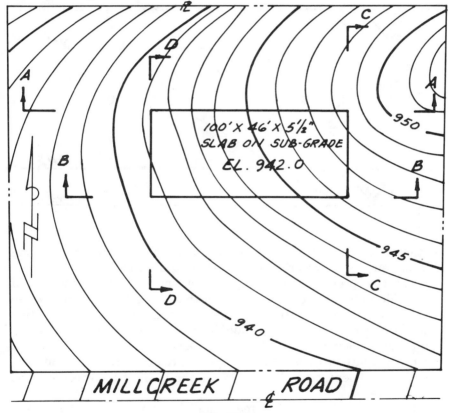

Fig. 10-1

Before drawing the finish grades, and from them determining the quantities involved, it is often helpful to plot a few cross-sections in order to get an overall picture of the project. Sections A-A and B-B (Figures 10-2 and 10-3) show the conditions along the north and south lines respectively of the slab area.

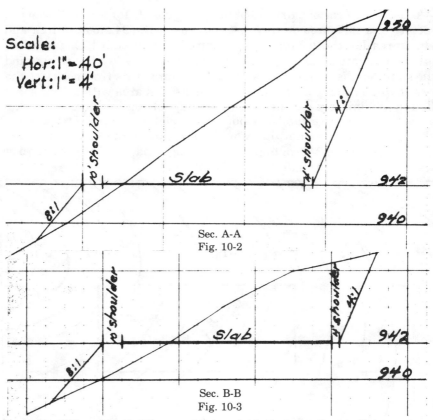

Scale:
Hor: 1" = 40'
Vert: 1" = 4'

Sec. A-A
Fig. 10-2

Sec. B-B
Fig. 10-3

Sections C-C and D-D (Figures 10-4 and 10-5) show the conditions along the east and west lines respectively of the slab area.

From these preliminary sketches, it is at once obvious that there will be considerably more cut than fill and this is the point in time when there should be consultation with the engineer as to whether, in this case, the elevation of the slab should be raised in order to obtain a more equitable balance of cut and fill, or, re-orient the slab to perhaps a northwest-southeast position which would require much less cut.

For the moment, we will continue on the basis of the specifications given above.

The outline of the shoulder areas should be drawn so that the cut and fill slopes can be developed from the outer limits of the planned level area. Since the fill slope is to be 8:1, that means that each drop of a one-foot finish contour will be horizontally eight feet out from the one above it. Conversely, for the cut slope of 4:1, it means that each rise of a one-foot finish contour will be horizontally four feet back from the one below it.

To show the *finish fill* contours on the drawing, set off a series of points at eight-foot intervals northerly, westerly and southerly from the outer limits of the shoulder areas and draw lines through these points,

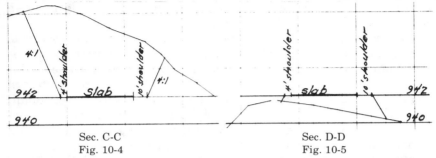

Sec. C-C
Fig. 10-4

Sec. D-D
Fig. 10-5

parallel with the north, west and south limits of the level area. Join each group together to show each *finish* contour elevation successively downhill except any portion which lies ABOVE its corresponding *natural* contour. This procedure is shown in Figure 10-6 and you can see

Fig. 10-6

how the 938 *finish* contour lies above the natural ground in the northwest part, meets its own level and follows that on the west and then has a
small portion of fill at its southwest corner. Because you could not from a
practical standpoint "square-off" the corners of the fill dirt, introduce a
round corner of about 6-10 foot radius.

To show the *finish cut* contours on the drawing, set off a series of
points at four-foot intervals northerly, easterly and southerly from the
outer limits of the shoulder areas and draw lines through these points,
parallel with the north, east and south limits of the level area. Join each
group together to show each *finish* contour elevation successively uphill
except any portion which lies BELOW its corresponding *natural* contour as shown in Figure 10-6. As with the finish grade lines for fill,
introduce a round corner of about 6-10 foot radius.

By connecting the intersections of the finish contour with the natural
contour with a dash line, you have the outline of the cut area and fill
area.

Consideration of Ingress and Egress.

Now consider the requirement for a driveway at approximately 5%
grade. Regarding slopes, the discussion thus far has dealt with a fraction
or proportion which relates the horizontal to the vertical measurement.
In other words, the first number (the numerator) in the fraction 4/1
represents the horizontal distance while the second number (the denominator) represents the vertical distance. These numbers expressed
either way, by the fraction or the proportion 4:1, mean that for every
four *units* (feet, meters, inches, centimeters, etc.) of horizontal measurement, there is one *unit* (of the same designation) of vertical
measurement.

Grade Percentage

What is the relationship of a percent of grade to slopes expressed as
above? Percentage is usually based on 100 units so a 5% grade, for
example, means that for every 100 feet (metres etc.) of horizontal distance, the slope is raised, or lowered, five feet (metres etc.) which in
other terms would be 100/5 or 100:5 or, reduced to its lowest common
denominator, 20/1 or 20:1. The *angle* of slope is seldom used but it is
derived on the basis of the tangent of the angle which of course is
obtained by dividing the amount of vertical rise, or drop, by the horizontal distance. In the case of this 5% grade, the angle would be the tangent
of 5/100 or 2° 51' 45".

An easy way to remember this relationship is that a 1:1 slope equals a
45° angle and in terms of percentage, it would be a 100% grade. For a
more practical application, an 8% grade is the same as a 100:8 slope
which equals 12½:1 or an angle of 4° 34' 26". From another approach, a
5:1 slope equals 100:20 which equals a 20% grade or an angle of
11° 18' 36".

Turning back to the matter of the driveway, it appears that inasmuch as there is almost 20 feet at right angles between each of the contours from the south shoulder southwesterly down to Millcreek Road, you have practically a 5% natural grade already in existence. However, the 10-foot shoulder is not wide enough for turning around nor for conveniently parking several cars or trucks. Is there an area that can be developed for parking?

Looking on the westerly side of the slab and realizing that there will be an excess of dirt from the cut, you might consider filling that area to a level suitable for parking except for the fact that it would require an acute angle at the top of the driveway for entrance to it.

Daylighting

Let us consider the area adjacent to the southeast corner of the slab. If, instead of having the cut follow around the slab and shoulder area, the finish-grade contours were "daylighted"* to the south, it would open a triangular space on natural ground that would be suitable for parking and turn around. To study this matter, make a rough cross-section at

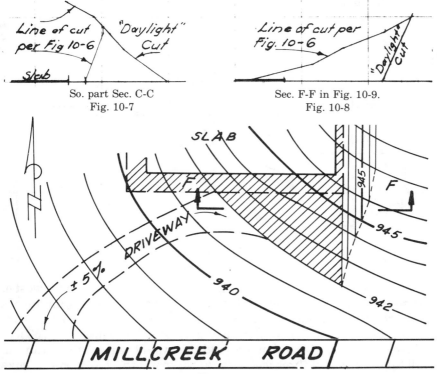

So. part Sec. C-C
Fig. 10-7

Sec. F-F in Fig. 10-9.
Fig. 10-8

Fig. 10-9

*To "daylight" is to project the cut plane to its intersection with the natural ground surface.

C-C with the finish contours "daylighted" as shown in Figure 10-7. Make another one on Section F-F as shown in Figure 10-8 taken from the revised plan drawing in Figure 10-9.

The hachured level area would then be easily accessible from the driveway and would provide parking for two or three cars plus turn-around area.

Is this the most feasible way? Because it would produce a still greater excess of cut dirt, let us reconsider the idea of parking on fill dirt on the west. By changing the direction of the driveway to follow the 940-foot contour, two advantages are accomplished: 1 — a minimal grade, and 2 — an easy and open access.

Again let us look at a rough cross-section through the westerly portion of B-B (Figure 10-6) and an extension of it through the west property line as shown in Figure 10-10. From this we see that by erecting a retaining wall along that west line and using a 4:1 slope, there is room for a proportionately greater amount of fill which in turn can provide a reasonable parking area.

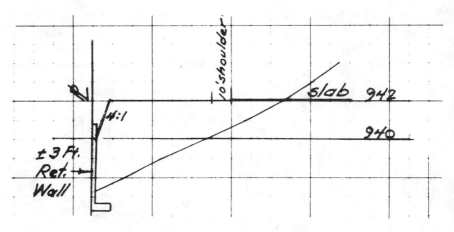

Fig. 10-10

Proceeding to the next step in this planning, redesign the area west of the slab using a retaining wall, the top of which will be nine inches above the finish grade. From the cross-section in Figure 10-10, it appears that a 4:1 slope up from the 940 contour will produce about 53 feet of level area west of the shoulder and this would require the retaining wall to be only about three feet high. At this point the development looks encouraging so let us also draw in the 4:1 slopes on the north and south sides of this new level area. See Figure 10-11.

After the earthwork calculations are made, you will know what the balance is between the cut and fill. If more dirt is needed for the parking lot, then that area south of the slab can be "daylighted" to furnish more fill.

Driveway to Road.

The first proposal for a driveway was from the southwest corner of the property in a northeasterly direction toward the building area which resulted in an approximately 5% grade; however, from further analysis, it would appear that by changing the driveway to a northwesterly direction, it could follow generally along the 940 contour and have a net difference of only two feet in elevation between its contact with the road and the parking lot level. This would result in about a 1.7% grade making it very easy to negotiate.

In the development of the previous project, the specification for the driveway was simple and the final location required a minimal amount of work. Many projects though require additional considerations such as banking one side and/or supporting the other side, super-elevating curves, limiting horizontal curves and/or introducing vertical curves.

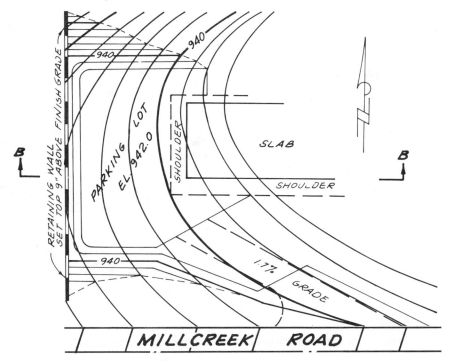

Fig. 10-11

EXERCISE 10-1: Take the drawing you made from the Annotated Field Notes in Figure 1-32 and superimpose the outline of the house and garage with driveway as shown on the sketch below using the measurements given. Set the elevation of the ground under the house at 115′ and under the garage at 115.5′. Slope the driveway down to the street holding controls as shown. Using a 4:1 slope for all cut areas on the north and east sides of the house, draw finish grades around the house leaving a 6 foot walkway next to the house at the elevations shown. It is suggested you make preliminary cross-sections on the north and east sides of the house and garage to help you see the picture before drawing the finish grade lines.

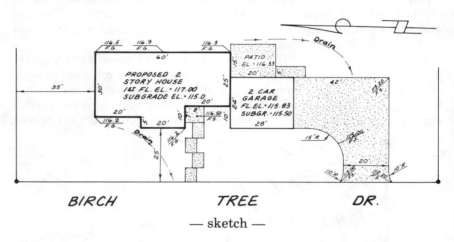

BIRCH   TREE   DR.

— sketch —

PROBLEMS

1. What is the main difference between plans and profiles for a R/W, and development plans for a building?

2. When both a boundary survey and a topographic survey are to be made on the same property, does it matter which is first? Why?

3. Is it necessary to plot cross-sections for a building site?

4. Would you do this before or after designing the finish grade? Why?

5. How do you designate the "cut" area and the "fill" area?

6. Which direction of the slope is represented by the numerator of the fraction, or the first number in the proportion, given for the slope designation?

7. What is the relationship between a per cent of grade and the slope?

8. Should you always accept the first proposal for a grading plan?

Chapter 11

Vertical Curves

Comparisons

Normally each horizontal curve for a road uses a single length radius and even when design based on the terrain requires a combination of curves with different radii in the form of compound or reverse curves, each part still has its single length radius. The vertical curve, in contrast, is based on the parabola which is not a single length radius. It is the most commonly used type of curve because it permits a gradual change in vertical direction and is more easily computed than a regular circular curve connecting two grades.

There are two types, namely: the *sag* and the *summit*. As illustrated in Figure 11-1, the summit curve lies below the tangent lines and is therefore concave downward while the sag curve lies above the tangent lines and is concave upward.

The rate of change of grade and sight distance are important controlling factors in designing a vertical curve.

Because of the physical demands of trains, railroads incorporate the flattest and longest vertical curves; next, the high-speed highways use less flat curves, and hilly roads or private drives may use shorter and sharper vertical curves. As a matter of safety, two items in particular

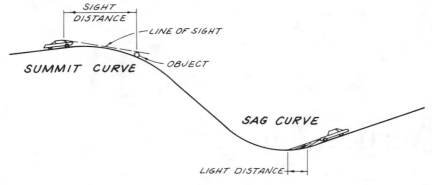

Fig. 11-1

must be considered in the design of a vertical curve in length and depth; namely: the speed of travel and the sight distance. With high speed, the transition from the constant grade into and out of the vertical curve must be gradual; with lesser speeds, the transition proportionally is less critical. At all times, however, consideration must be given to a line-of-sight distance that will assure safety.

Webster's definition of a parabola is: "a conic section, the intersection of a cone with a plane parallel to its side; a plane curve, any point of which is equidistant from a fixed point, the *focus,* and a fixed straight line, the *directrix.*" Figure 11-2 shows how the short cone creates a relatively shallow flattened curve while the tall cone makes a narrower sharper curve so there is an infinite number of forms a parabolic curve may have.

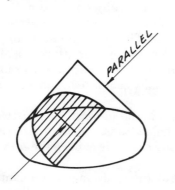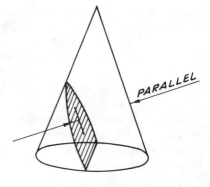

Fig. 11-2

The pattern of equidistant points which form a parabolic curve between a *focus* and a *directrix* is shown in Figure 11-3.

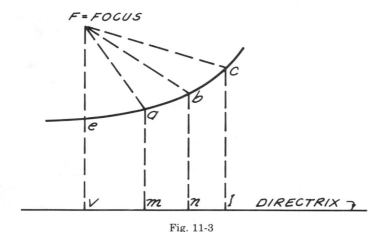

Fig. 11-3

Terminology

The elements of the vertical curve consist of:

L = the length of curve in stations *measured horizontally.*
g_1 = the initial grade, also called *back* tangent grade.
g_2 = the final grade, also called *forward* tangent grade.
r = rate of change of grade per station.

Grade is expressed as a percentage which is a term representing the units (feet or metres) of rise or fall per 100 units (feet or metres) horizontally. Review Page 10.6.

An uphill slope is called plus and a downhill slope is called minus. The rate of change of grade is the *algebraic* difference of the back and forward tangents, i.e.: $+4\% -(-3\%) = +7\%$ and $-5\% -(+2\%) = -7\%$.

Similarly to the horizontal curve, you establish and use the PI (Point of Intersection) of the two tangents of the parabolic curve. The intersection of the back tangent (which is the one approaching the vertical curve) with the forward tangent (which is the one receding from the vertical curve) is designated as VI or PVI. The beginning of the vertical curve is BVC and the end of it is EVC.

Parabolic curves may have *either* equal or unequal tangents.

The length of the arc of the curve between stations and the horizontal distance between stations are considered to be the same as far as their application to computing a vertical curve is concerned. For the purpose of computations and establishing ground control points, the curve is divided into equal segments called either *stations* or *stages*. In Figure 11-4 showing the parts and their relationships, S_1 = BR = BR_1, etc. While railroads normally use 100 foot stations, highways and roads may also use 100 foot stations on straight sections with a constant grade but will use smaller segments in order to get smoother transitions on the curves. This in turn also permits tighter control on the physical grading.

Vertical Curve Parts and Relationships.

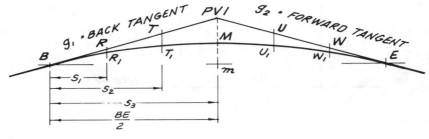

Fig. 11-4

There are two methods for establishing the final elevation for a station on the curve: 1. Points on the curve are calculated directly from the algebraic equation for a parabola; 2. Offsets are calculated (up or down) to add to, or subtract from, the same station on the tangent. In other words, for a summit curve, the offset is subtracted from the tangent but for a sag curve, the offset is added; consequently, the tangent elevations at each station must be established first and then the curve elevations at the respective stations are determined by the vertical offset relationship.

The elevations at the stations on the tangent are obtained by multiplying the grade percentage by the distance from the terminal point, either the BVC or the EVC of the curve, and adding that figure to the terminal point elevation for a summit curve or subtracting that figure from the terminal point elevation for a sag curve.

Formulas and Their Application

First we will consider the *symmetrical curve* which means that even though the elevations at the BVC and EVC may be different, the tangents are equal.

The theorem for establishing elevations at points on a parabolic curve by tangent offsets from grade lines is based on three properties of an equal-tangent vertical curve:

1. The offset points along the two grade lines are symmetrical with respect to their intersection, the PVI, regardless of whether the elevations at the BVC and EVC are identical or not.
2. The offsets from the tangent to the curve at any point are proportional to the squares of the horizontal distances from the point.
3. The curve itself lies midway between the vertex of the grade lines, the PVI, and the midpoint of the long chord BE. In other words, VM = Mm in Figure 11-4.

The first formula is, using the terminology on page 11.3,:

$$L = \frac{g_2 - g_1}{r}$$

where r has been specified. If the length, L, is given or assumed, then r must be computed.

Vertical Curve by Equation

The equation for the parabola to connect two grades based on the X and Y axis as shown in Figure 11-5 is: $Y = ax^2 + bx + c$

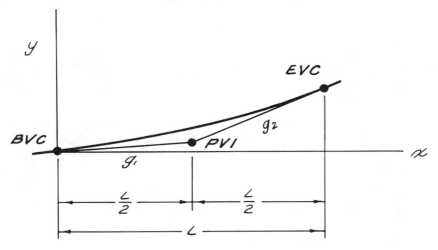

A parabolic curve related to X and Y axis.
Fig. 11-5

Starting with measurements from 0 on the X axis, Y = the elevation of a point on the curve (c is 0 at the Y axis) and the slope of the curve is g, so $b = g$. Incorporating also the derivative of the rate of change of grade, r, the result is:

$$Y = \frac{r}{2} x^2 + g_1 x + c$$

Of course, r must be given its correct algebraic sign. The easy way to remember it is that for summit curves, that is a curve concave downward, the sign is *minus* and for sag curves, a curve concave upward, the sign is *plus*.

The following example illustrated in Figure 11-6 shows the application of the formula.

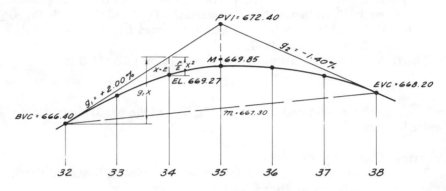

Parabolic vertical curve developed from formula.
Fig. 11-6

The length of the curve, L, has been determined to be 600 feet and the elevation at the intersection of the tangents, PVI, is 672.40. With the gradients given at +2.00% for g_1 and −1.40% for g_2, calculate the elevations at the BVC, EVC and the 100-foot stations between them.

Solution: The elevation at the BVC will be obtained by subtracting ($g_1 \times 3$ stations) from elevation 672.40 at the PVI and the elevation at the EVC will be obtained by subtracting ($g_2 \times 3$ stations) from elevation 672.40 at the PVI.

    Elev. BVC = $672.40 - 2.00 \times 3 = 666.40$ ft.
    Elev. EVC = $672.40 - 1.40 \times 3 = 668.20$ ft.

The change of grade per station is

$$r = \frac{g_2 - g_1}{L}$$
$$= \frac{-1.40 - (+2.00)}{6} = -0.567\%$$

The point elevations that establish the position of the curve at each station are derived from the equation:

$$Y = \frac{r}{2} x^2 + g_1 x + c$$
$$= -0.283 x^2 + 2.00 x + 666.40$$

The following table shows the values thus obtained.

Station	x	x^2	$\dfrac{r}{2}x^2$	$g_1 x$	Elev. BVC	Elev. Pt. on curve
BVC = 32	0	0	0	0	666.40	666.40
33	1	1	− 0.28	+ 2.00	666.40	668.12
34	2	4	− 1.13	+ 4.00	666.40	669.27
35	3	9	− 2.55	+ 6.00	666.40	669.85
36	4	16	− 4.53	+ 8.00	666.40	669.87
37	5	25	− 7.08	+10.00	666.40	669.32
EVC = 38	6	36	−10.19	+12.00	666.40	668.21

Note that the final point elevation is the sum of the three preceding columns. Because of "rounding off" hundredths, the EVC elevation shows 0.01 more than in Figure 11-6. The 668.20 would be held.

By taking the average of the elevations at BVC and EVC, you get the elevation at m, then the average of the elevations at m and PVI gives the elevation at M which checks the elevation at that point obtained from the application of the equation. Besides showing the check-out of the elevation for M in Figure 11-6, you can also see the graphic representation of the algebraic sum of the figures in columns 3, 4 and 5 for station $x = 2$.

Vertical Curve by Tangent Offsets

Relative to producing a vertical curve by tangent offsets from grade lines, the second property in the theorem is illustrated in Figure 11-7 where RR_1, TT_1, etc. is parallel the Y axis and S_1, S_2, etc. is parallel the X axis.

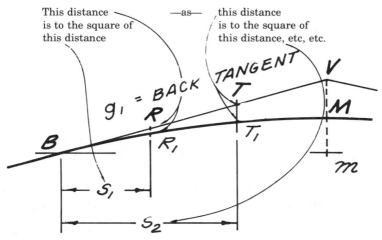

This distance is to the square of this distance —as— this distance is to the square of this distance, etc, etc.

Illustration of "second property"
Fig. 11-7

Therefore: $RR_1 : (S_1)^2 = TT_1 : (S_2)^2$

If "d" represents the difference, or vertical distance to the slope offset, between the tangent line BV and the curve BM at 1 station (100 feet) from the beginning of the curve (B), and S_1 is the horizontal distance in station from B to a point R_1 on the curve, then the vertical offset distance will be expressed by the equation:

$$RR_1 : (S_1)^2 = d:1^2$$
$$\text{or} \quad RR_1 = d(S_1)^2$$

We will apply this formula to the conditions shown in Figure 11-8.

Points on the vertical curve for specific elevations are usually set in stages of 100 feet, 50 feet, 25 feet, 20 feet (50 metres, 20 metres, 10 metres), or whatever amount will suit the interval change required.

If the difference of elevation is small, the control points will be satisfactory at 100-foot intervals. If the change of grade is rapid, the points should be closer together in order to give a smoother transition control.

The total length of the curve, usually in stages of 100 feet, is based on the permissible *change in the rate of grade* established by the governing body and the required *sight distance*.

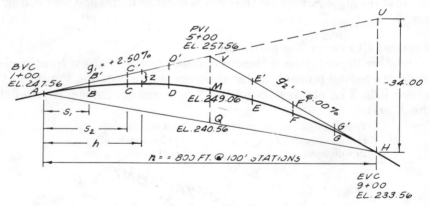

Facts for example of a symmetrical curve.
Fig. 11-8

From the first property (Page 11.5), the offsets B'B = G'G, C'C = F'F and D'D = E'E. By the second property, C'C = 4 B'B, D'D = 9 B'B, VM = 16 B'B = 16 G'G, E'E = 9 G'G and F'F = 4 G'G. By the third property, M is the mean elevation between V and Q so VM = M Q and $Q = \dfrac{A + H}{2}$. As before, the elevations at the BVC and EVC are derived from the elevation at the PVI minus the grade times the length of the tangent. Calculate the curve point elevations for all stations from the BVC to the EVC.

Solution: Since the offsets at B′, C′, D′, E′, F′ and G′ are proportional to the offset at V, the PVI, as their respective singular distance is to the total distance from BVC to PVI and PVI to EVC, the S distance in the formula can be set forth as a fraction of VM.

The elevation at Q is the average of the elevations of A and H and the elevation at M is the average of the elevations at V and Q. From Figure 11-8 the elevation differential between V and M is 8.50 feet which will be the basis for developing the fractional vertical offsets of d, page 11.8.

In order to apply this offset, however, it is necessary first to develop the elevation for each station on each tangent.

Profile Station	Prob. Sta.	Elev. on Tangent	S	S²	Vert. Offset	Elev. on Curve	1st Diff.	2nd Diff.
1+00	A	247.56	0	0	0	247.56		
							−1.97	
2+00	B′	250.06	1/4	1/16	−0.53	249.53		1.07
							−0.90	
3+00	C′	252.56	2/4	1/4	−2.13	250.43		1.05
							+0.15	
4+00	D′	255.06	3/4	9/16	−4.78	250.28		1.07
							+1.22	
5+00	V	257.56	1	1	−8.50	249.06		1.06
							+2.28	
6+00	E′	251.56	3/4	9/16	−4.78	246.78		1.07
							+3.35	
7+00	F′	245.56	2/4	1/4	−2.13	243.43		1.05
							+4.40	
8+00	G′	239.56	1/4	1/16	−0.53	239.03		1.07
							+5.47	
9+00	H	233.56	0	0	0	233.56		

The "1st Difference" and "2nd Difference" columns in the above table give you a check on your mathematical development of the curve elevations. The "1st Difference" is, as implied, the vertical relationship between succeeding points on the curve; therefore, if the "2nd differences" between those first differences is within two or three hundredths of a foot of each other, then you have a reasonable constant and an assurance of the correctness of your point derivation.

Intermediate Points and Clearances.

On occasion, problems arise as to whether there is sufficient overhead clearance for trucks at a certain spot on the vertical summit curve or is there enough fill over a pipe crossing at a particular location on a sag curve to properly cushion the load on the pipe, etc., etc. These questionable areas are usually not at a summit point or at a station previously calculated but at some intermediate point, in which case, the required station and elevation point on the vertical curve must be computed.

Use the same curve equation as given on Page 11.6:

$$Y = \frac{r}{2}x^2 + g_1\,x + c,$$

where x will be the desired odd station point beyond the BVC, c will be the elevation at BVC and y will become the elevation at the desired point.

This can also be accomplished by using the tangent offset method which is a little less direct. Again the elevation of the desired station on the tangent line is obtained by taking the gradient times the distance from the BVC, or back from the EVC if it is in the forward tangent. At this point, the offset is figured on the proportional basis discussed with, and shown in, Figure 11-7 using the formula $RR_1 = d(S_1)^2$.

Summit Point

In order to determine the clearance in a tunnel or under a bridge, it is necessary to establish the highest point, known as a Summit Point or Apex, on a vertical curve. It is also the position on the curve where a line tangent to it has a grade of 0%. See Figure 11-9. Conversely, it is necessary to determine the lowest point, known as an Inverse Summit Point, on a sag curve in order to know where water will cross or settle on the roadway or by which to establish elevations for sub-structures under the road bed. Likewise it is that point on the vertical curve where a line tangent to it has a grade of 0%. These "Summit Points" are seldom at the midpoint of the curve.

The Summit Point or Apex of a curve is the point
at which the grade of the curve is zero.

Fig. 11-9

Without going through the algebraic development, the following equation is used for locating the *station* at which point the line of tangency equals 0% grade. Referring to the facts given in Figure 11-8:

$$h = g_1\left(\frac{n}{g_1 - g_2}\right)$$

so: $$h = +0.025\left(\frac{800}{0.025 - (-0.06)}\right) = 235.29$$

therefore: BVC Station   1+00
                    + h   2+35.29

              = Station   3+35.29 for point of zero
                                 grade tangency.

The equation for computing the *elevation* at 0% grade is, using the facts shown in Figure 11-8:

$$Z = (\frac{h}{n})^2 \text{ UH}$$

$$Z = \left(\frac{235.29}{800}\right)^2 (-34.00)$$

$$= -2.94$$

therefore: elevation of Summit Point on curve at Z
              = elev. at BVC + (335. g_1) − Z
              = 247.56 + 8.38 − 2.94
              = 253.00

Asymmetrical Curves

You have seen with the simple basic symmetrical curve, with only a few stations involved, that it is easy to use the fractional part method to compute the offsets for elevations on the curve, but when a long curve with many stations is involved, the fraction method is not as practical as the pure equation.

The singular characteristic of the vertical asymmetric curve is that it has unequal curve-tangent lengths. In other words, the distance between the BVC and the PVI is different from that between the PVI and the EVC. See Figure 11-10.

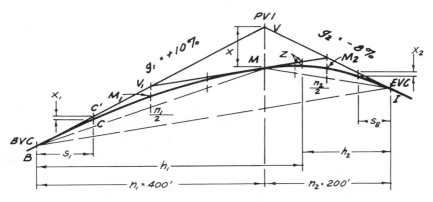

In an asymmetrical vertical curve,
the tangent lengths are unequal.

Fig. 11-10

In fact, the asymmetrical curve has two separate parabolas which are tangent to each other at the vertical axis; therefore, each part is to be solved on the same basis as an individual symmetrical curve.

Looking at Figure 11-10, you see that the parabolic curve on each side of the vertical axis through the PVI is similar to the single symmetrical whole curve as shown in Figures 11-6 and 11-8. Each side, or each symmetrical curve, has the midordinate on the vertical axis (M_1 and M_2). Each is divided into even stations, etc.

Again, the fundamental rule applies to each parabola – that the vertical offset from the tangent to the curve is proportional to the square of the horizontal distance from the point of tangency to that of the vertical offset.

For the long curve on the left in Figure 11-10:

$$\frac{VM}{C'C} = \frac{n_1{}^2}{s_1{}^2}$$

or $\quad \dfrac{X}{X_1} = \dfrac{n_1{}^2}{s_1{}^2}$

or $\quad X_1 = X\left(\dfrac{s_1{}^2}{n_1{}^2}\right)$

where: X = distance of offset on vertical axis at PVI in feet from tangent intersection to curve.

X_1 = distance of vertical offset in feet from tangent line to any point on the curve between BVC and PVI.

n_1 = horizontal distance from BVC to PVI in 100 foot stations.

s_1 = horizontal distance from BVC to vertical offset X_1 in 100 foot stations.

For the short curve on the right in Figure 11-10:

$$\frac{X}{X_2} = \frac{n_2{}^2}{s_8{}^2}$$

or $\quad X_2 = X\left(\dfrac{s_8{}^2}{n_2{}^2}\right)$

where: X = same as for long curve.

X_2 = distance of vertical offset in feet from tangent line to any point on the curve between PVI and EVC.

n_2 = horizontal distance from PVI to EVC in 100 foot stations.

s_8 = horizontal distance from EVC to vertical offset X_2 in 100 foot stations.

Note: The application of calculated second differences for the purpose of checking elevations on a symmetrical curve is not adaptable to the asymmetrical curve as a whole but may be used on each of the two parabolas separately.

On occasion the drawing of a vertical curve can be done with circular curves. One procedure for obtaining the radii is by the use of the nomograph shown in Figure 11-11. The method of using it is as follows:

1. Compute the algebraic difference in grades.

2. If necessary, evaluate the expression: $S = \dfrac{(100)^2}{(H)} V.$

3. Place straightedge through algebraic difference of grades on "G_1 - G_2" scale and length of vertical curve on "L" scale and intersect turning line.

4. Turn through point on turning line to value on S as determined in Step 2. (This will remain a constant for a given set of scales).

5. Read length of radius on R.

RADII FOR DRAWING VERTICAL CURVES WITH CIRCULAR CURVES

Equation: $R = \left(\dfrac{100}{H}\right)^2 \dfrac{L V}{(G_1 - G_2)}$

R = RADIUS OF CIRCULAR CURVE, INCHES
L = LENGTH OF V.C., STATIONS
H = HORIZONTAL SCALE (FT. PER IN.)
V = VERTICAL SCALE (FT. PER IN.)

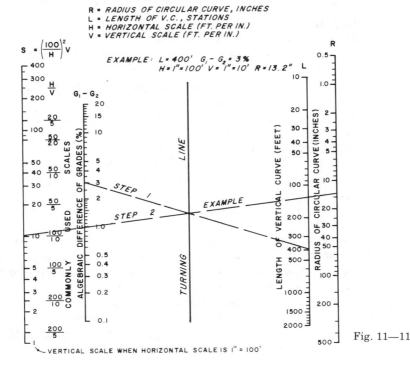

Fig. 11—11

The profile should be examined for these two limiting factors:

a. Long vertical curves having relatively small radii. These often cannot be drawn with a simple circular curve. They may be drawn with a french curve or as a compound circular curve.

b. Large differences in elevation between the two ends of the curve, usually, caused by curves joining two steep grades in the same direction. The curve indicated by the nomograph will be too small. The correct curve can be estimated by the proportion:

$$\frac{\text{calculated radius}}{\text{correct radius}} = \frac{\text{horizontal curve length (inches)}}{\text{scaled distance between curve ends (inches)}}$$

EXERCISE 11-1: Develop the elevations, vertical offsets and elevations on the curve for a symmetrical curve from the following facts:

Elevation at PI	= 788.32′
Initial grade	= +3.40%
Final grade	= −6.60%
BVC	= Sta. 6+00
EVC	= Sta. 14+00

Tabulate your figures in the same form as shown on page 11.9 using the first and second differences to check your work.

PROBLEMS.

1. Describe the two types of vertical curves.
2. What are the two major controlling factors in designing a vertical curve?
3. Describe the differences between the principles of the horizontal curve and the vertical curve.
4. Describe the simplest way to *draw* a vertical curve based upon your description above.
5. What are the *elements* of the vertical curve?
6. Illustrate the "rate of change" of a grade with an example.
7. What is the relationship of the distance along the curve and along the tangent between stations?
8. Name the two methods for establishing station elevations on a vertical curve.
9. Describe the three properties basic to the theorem used for developing station elevations on a vertical curve by tangent offsets.
10. Explain the easy way to remember the algebraic sign for r in the equation for the rate of change of grade.
11. When should you use stations close together?
12. Give a parallel description of a "summit point".
13. Name the two conditions that distinguish an asymmetrical vertical curve from a symmetrical one.

Chapter **12**

Special Type Maps

Special Type Maps

There are numerous types of maps upon which the draftsman is called to create from facts existing in public or private records or perhaps only in the thoughts or imagination of a person. Here we will consider some that are common and a few not so common. Maps for the subdivision of land are known as subdivision maps or tract maps or parcel maps. Occasionally a map is attached to a document and is thereby called a deed map. A leasehold map is made to illustrate the areas covered by leases in a shopping center, an industrial complex, oil drilling areas, etc.

When a roadway is to be moved to a new location, it is desirable to have a map showing the existing right-of-way and its relationship to the new right-of-way delineating the boundary control and measurements as shown in Figures 1-23 and 24. Although the above shows a limited area, it may well be a small part of a large strip or right-of-way map. As usual, the job conditions govern the requirements imposed on the map; it might be simple or complex.

The special map required by title insurers for the issuance of a policy in conformity with the requirements of the American Land Title Association is known as an A.L.T.A. Survey map. This becomes the most detailed map you will experience.

Another very particular type of map is that used in court cases, the effectiveness of which is determined by its illustrative impact.

Visibility profiles are used by the Forest Service to position their lookout towers at the most advantageous points. This type of study is also used for checking microwave transmission installations.

Part Ownerships

Several methods are used for the conveyance of subdivided land: parcels or lots shown on a map commonly called a tract map, with singular identifiers assigned to each parcel or lot; individual metes and bounds descriptions of parcels; a combination which uses a metes and bounds description with a map attached, sometimes called a deed map, showing the particular piece of land; or a map of the parcel, or parcels, not made in the form, nor fulfilling the requirements, of a statutory tract map, attached to the document of conveyance and whereby reference is made in the document to a particular parcel shown on the map.

Although a large area of land may have been subdivided into a pattern of parcels such as a town subdivision or sections in a township, many times those parcels are cut again into any number of smaller parts resulting in separate ownerships.

In the case of section land, the cuts are often made as a half, a quarter, a quarter quarter, etc., or by metes and bounds, and subdivision parcels shown on a record map may also be cut by either method. In any situation, these ownerships of parts of parcels must be delineated on a map. This in effect is what you see on assessor parcel maps, one of which is reproduced in Figure 12-1.

Subdivision-of-Land Maps

Looking at the final map of a small 20-lot subdivision as shown on Page 12.6, recorded in the official book of maps in the office of public records, it does not appear complicated. To understand the background, however, let us review the step-by-step procedure used in not only creating the map itself but also in meeting the requirements of some of the public offices concerned with subdividing land.

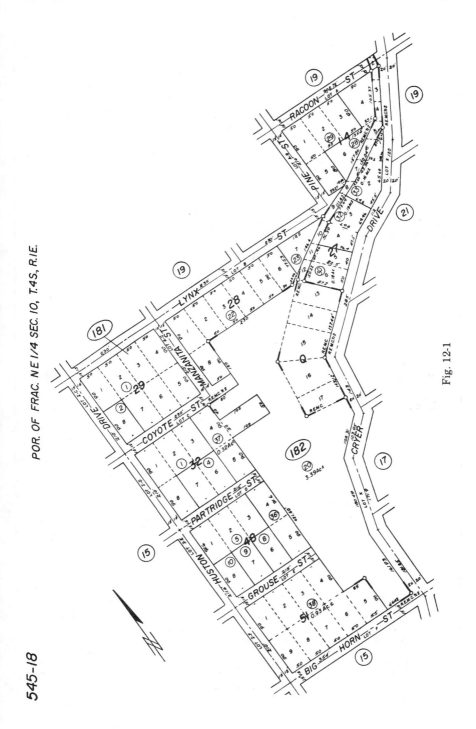

POR. OF FRAC. NE 1/4 SEC. 10, T.4S., R.IE.

Fig. 12-1

545-18

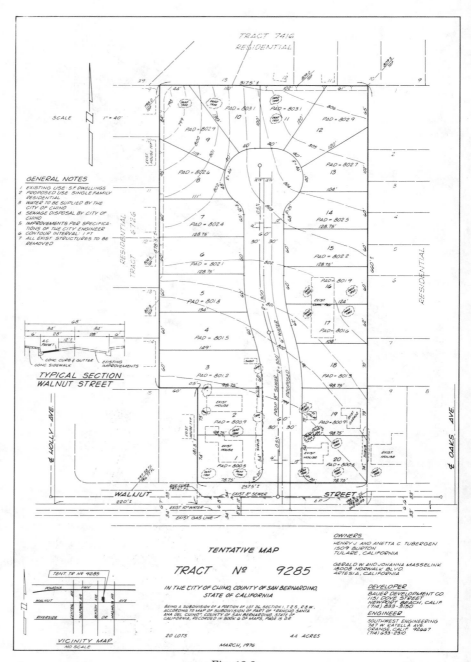

Fig. 12-2a

Prior and supplementary to the submission of the Final Map, other necessary operations are performed which the public offices never see. These include a review of the topography, study maps of the lot and street layout, drainage and sanitation conditions, preliminary surveys, review of the preliminary title report with consideration of easements and analysis of any boundary line problems.

Tentative Map

When an engineer or surveyor is given a legal description by his client and told to do the necessary work for the preparation and filing of a tract map, his first step is to notify the proper public office, such as the county or parish surveyor, of his intention by giving the location of the developer's land and asking for a new tract number or name by which to identify it. If a name is submitted, it is checked to be sure there is no duplication of previous tracts. In the case of a number, the next higher one over the last one issued will be assigned.

The Tentative Map as presented is seldom the first draft of the layout; often there are several "study maps" made first to consider the adaptability of the lot sizes and configurations to the topography of the land as well as to analyze any sewerage and drainage flow problems.

Following this, the engineer or surveyor has his office draw the outline of the property on a sheet to be known as a Tentative Map which is often, but not necessarily, the same size as the standard established for the final map. The one prepared for the final map shown in Figure 12-2b is shown in Figure 12-2a. Within the boundary lines are shown the contours and topography of the existing conditions on the land along with a proposed layout of lots, streets, alleys, walkways, parks and any special items to be developed.

A part of this depends upon the preliminary report by the title or abstract company from which all easements shown in the report are plotted on the map to determine if any will need to be relocated in order to accommodate the street and/or layout, or if, in consideration of the cost of changing the physical occupation of a pipe line, power line, etc. in an existing easement, it may be more economical to redesign the street and lot layout.

The final boundary lines of the Tract should now be established so that no changes will be made to affect the inside development. The title report along with boundary deeds and maps should be reviewed so as to be certain that the surveyed boundary and the record boundary are compatible and will be insurable. Of course all interior conditions must fit and be tied to the boundary lines.

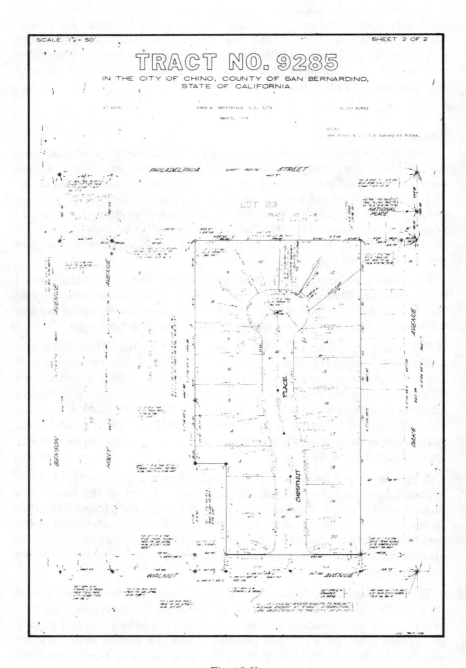

Fig. 12-2b

Whereas Figure 12-2b shows an excellent boundary study which is made a part of the Final Map, an irregular shaped area can be much more involved as shown in Figure 12-2c.

Sometimes studies of monumentation-on-the-ground problems must be blown up for analysis before making a final decision. Figure 12-2d illustrates such a situation.

Usually the approval of the Tentative Map by the governing body is made subject to conditions which they deem necessary in order to conform with the Master plan requirements previously made for the area under their jurisdiction. These may require some redesigning of the lots and street layout and/or drainage and sanitation facilities. In any event, when the ultimate layout has been determined, all calculations of lots, streets, alleys, walks etc and centerline control for the Final Map are now needed and a "Calculation Map" is in order. An example is shown in Figure 12-2e.

If a computer plotter is to be used, all the details and juxtaposition of the many items must be determined at this point in time.

Also on this same sheet is to appear the name of the owner and/or subdivider-developer, the name of the private engineer or surveyor, a description of the land (sometimes complete but usually only in a caption form), the tract name or number, an index or location map and the following items with either a short explanation of the status of the item or reference to documentary evidence or a letter concerning it: —

1. Water.	7. Power and street lights.
2. Sewer.	8. Gas.
3. Flood and drainage conditions.	9. Fire protection.
4. Street improvements.	10. Hospitals.
5. School facilities.	11. Transportation.
6. Parks and recreation.	12. Access to highways.

This Tentative Map is presented with the required number of copies to the clerk of the advisory agency, such as the Planning Commission, or if there is no such agency, with the clerk of the legislative body or such officer or employee as may be designated by local ordinance, to review the planned subdivision. (You should get your local ordinance to determine this procedure.) From a study of this map, decisions are made as to whether the proposed method of constructing the improvements are acceptable and/or if they are sufficient for the overall development. Sometimes adjoining or intervening areas are included in the consideration; for example, if the main line sewer and drainage facilities for the instant tract are to subsequently also serve land beyond this development, the governing body may require the size of pipe to be increased accordingly. Usually some financial arrangement is made for this excess requirement.

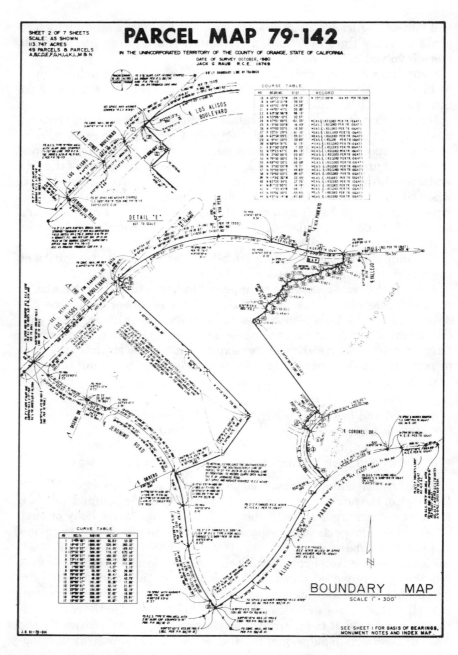

Boundary Map

Fig. 12-2c

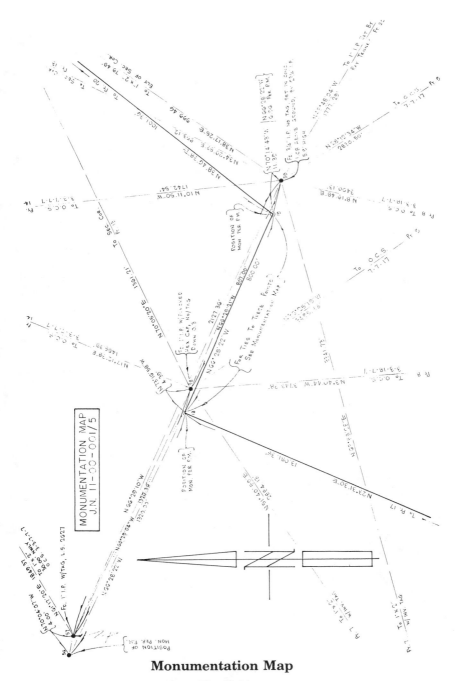

Monumentation Map

Fig. 12-2d

Calculation Map

Fig. 12-2e

Final Map

When the engineer or surveyor receives approval of his Tentative Map, along with any specific denials or requirements, he proceeds with work on the final map along with construction plans and profiles. If the size of the tract will permit, it can be included on the Title Sheet 1, otherwise it may take any number of sheets besides the Title Sheet.

A further check of the title report is made, this time for the names of all owners of interest in the land who are to appear on the title sheet so that the proper certificates can be made for the principals to sign, along with notary forms. In some areas the signatures of holders of easements are not required if the easement cannot ripen into a fee. The decision relative to such an omission is usually up to the political entity having jurisdiction over the area under consideration.

Certificates to be put on the title sheet of the tract map include, but are not limited to, the following:

1. Private engineer or surveyor.
2. Public engineer or surveyor.
3. Approval by city. (This is usually signed by the City Clerk confirming a formal resolution. If the tract is not within the corporate limits of a city at the time of recording but is within an area to be annexed to the city or in such close proximity that by agreement with the county [or parish], the city's approval is required by ordinance.)
4. Approval by county or parish. This likewise is often signed by the Clerk confirming a resolution by the governing body.
5. Confirmation by the Tax Assessor that either taxes have been paid (this covers the first half of the fiscal year immediately following the issuance of tax bills) or a bond has been posted (this covers the remainder of the tax year before the amount of taxes for the following year has been computed and released).

Certain statements of fact are also required on the title sheet, some of which are given specific forms, while others have none; these include. but are not limited to, the following:

1. Caption type (not a complete metes and bounds) description.
2. Basis of bearings.
3. Monument Note.
4. Bench Mark (if required as in case of a condominium).
5. Legend (if required).
6. Omission of certain signatures (if applicable).
7. Index map of subsequent sheets (if many).
8. North point and scale.
9. Soil test report (if required).
10. Any special "Note(s)" that may be pertinent to the project or requested by the political entity involved.

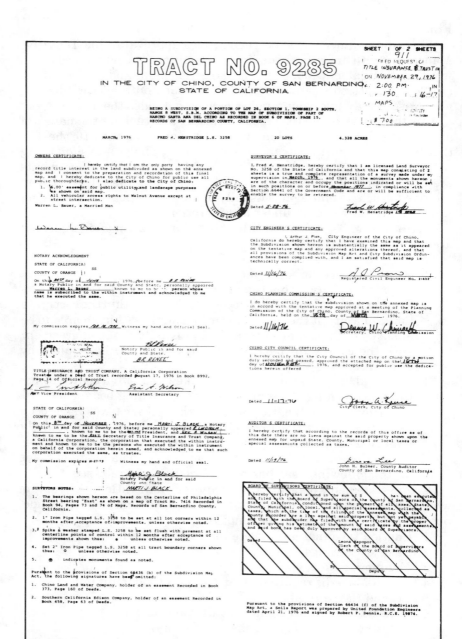

Fig. 12-4.

Figure 12-4 shows the title page for Tract No. 9285.

Pertaining to the inside of the final map, every segregation of an area shown, which includes lots, streets, parks, alleys, walkways, etc., should have bearings and distances given in such clarity and sufficiency that each area can be mathematically checked and therefore correctly surveyed and located on the ground.

The allowable error of closure for the various sized areas including the perimeter of the tract usually depends upon prevailing customary practice which in turn is based on the value of land and/or the requirements established by the County Surveyor or City Engineer who has jurisdiction over checking the map. In this matter of closure, for an example of land of high value, the perimeter of *each* area, lot, block, centerline of streets around a block, etc., may be required to close within 0.007 of a foot. In urban localities with lower value land, the required closure may be only 0.020 of a foot each area.

Each parcel on the map, but not including streets, alleys, walkways and the like, is to be given a unique identifier; this is usually in the form of numbers in consecutive order. Letters are sometimes used for special areas such as parks.

In some states, the subdivision of land is accomplished through the same general requirements as set forth above, but with maps and procedures called by different names.

Plans and Profiles.

With the boundary established, and streets laid out, work proceeds on the design of utilities and drainage with their appurtenant facilities. Plans and profiles take shape for each item required two of which are shown in Figures 12-5 and 6.

Some utilities planned for a new tract will require engineered plans and profiles while others will not. For example, water pipe lines and subgrade power and telephone conduits do not need careful grading control developed from plans and profiles. For these utilities, horizontal locations are shown on the map by easements over lots when necessary but such right-of-ways are not needed for any part in streets, alleys or walkways because they are customarily dedicated for public use. These are often installed at whatever constant depth can be achieved by a mechanical ditch digger without interfering with other lines.

Sanitation

Sewer facilities, whether in easements over the rear of the lots or in the streets or alleys, must have plans and profiles for the overall control of the system within the tract and its relation to external main lines.

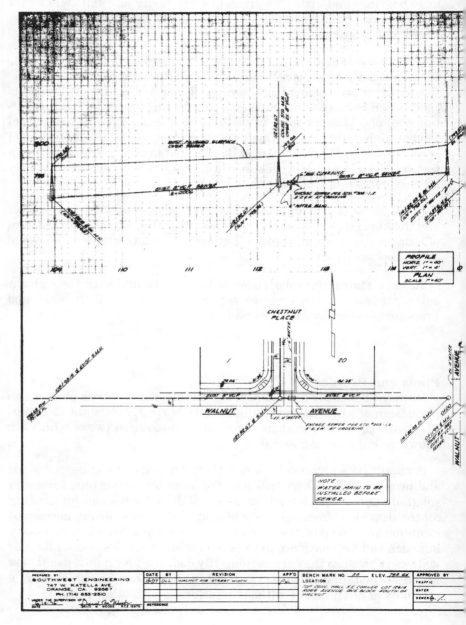

This utility may be separate or combined with the street work depending on the complexity of the overall picture and the requirements of the local governing authority. The one for this Tract is shown in Figure 12-5.

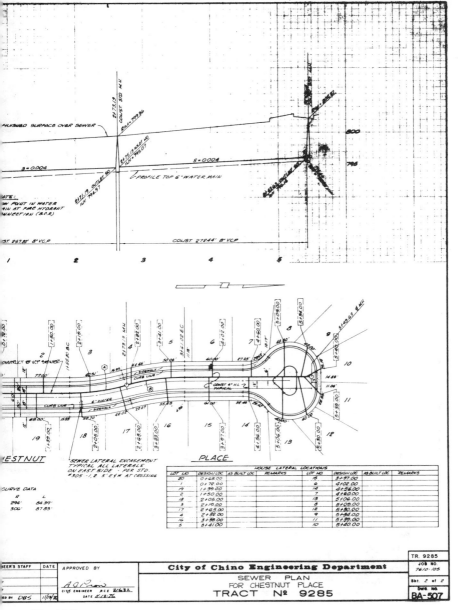

Fig. 12-5

Streets

Extensive plan and profile work is concerned with the street improvements, an example of which is shown for this Tract in Figure 12-6.

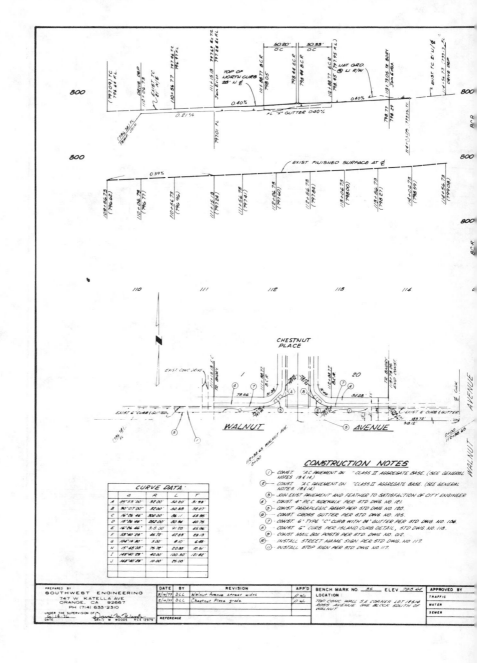

Drainage

Whether there is little or great slope to the land, drainage must be
considered; depending upon the circumstances, it may be accomplished

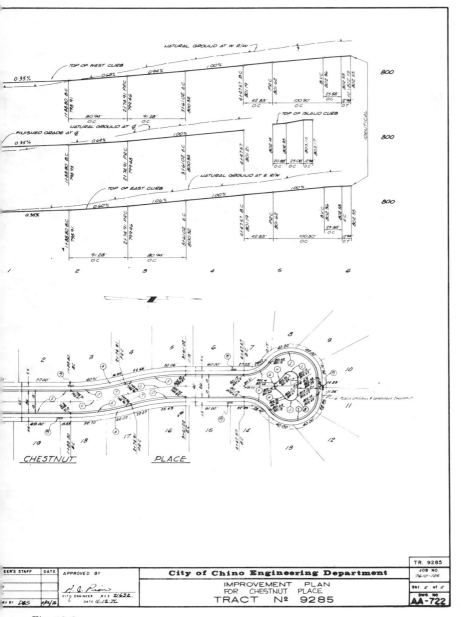

Fig. 12-6

all on the surface or with some underground structures to help. This may all be included within the street work project unless it is very complicated and needs separate delineation. For this Tract 9285, some is

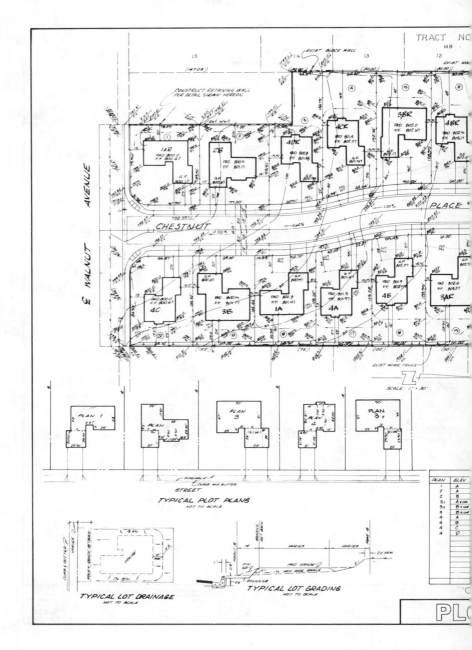

shown on the Street Plan and Profile (Fig. 12-6) and the remainder is shown on the Plot and Grading Plan reproduced in Figure 12-7.

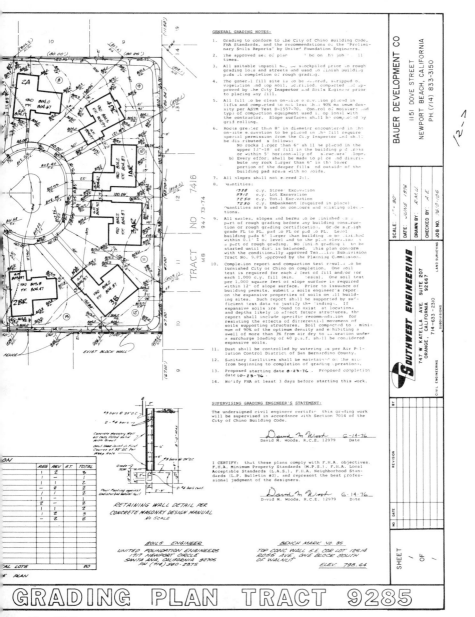

Fig. 12-7

Plot and Grading Plan

This must be very comprehensive and include not only the street area but also the positioning of the buildings as shown in Figure 12-7.

EXERCISE 12-1: Now that we have followed the matter of developing a new subdivision of land through its various phases of tentative map and final map with plans and profiles, some of which have been further discussed in Chapters 6 through 10, take the following description of a developer's parcel of land and the field notes on the survey of that land and draw your own tentative map and final map.

Description: Lots 3, 4 & 5 in Tract No. 504 in the City of Costa Mesa, County of Orange, State of California, as per map recorded in Book 17, Page 31, of Miscellaneous Maps, records of said County. EXCEPT the southwesterly 137.50 feet of said Lot 5.

Following are the requirements made by the City for the development of this Tract:
1. Each lot shall have a minimum of 6,000 square feet.
2. 50 feet is the minimum allowable width.
3. Corner lot must be 60 feet wide and have a 17 foot property line radius at the street intersection.
4. 20 feet must be dedicated and improved for widening 19th. Street on the northeast.
5. 10 feet on the southwest must be dedicated and improved for one-half of a 20 foot alley. The south line of the property will coincide with existing sewer line.

Also shown here are the typical cross-sections required by the City.

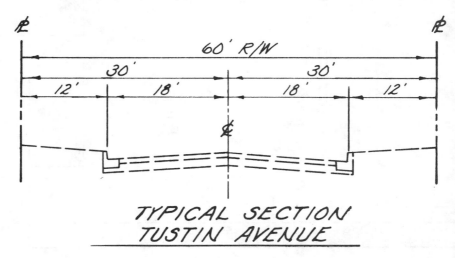

TYPICAL SECTION
TUSTIN AVENUE

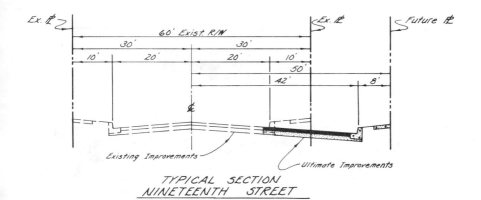

TYPICAL SECTION
NINETEENTH STREET

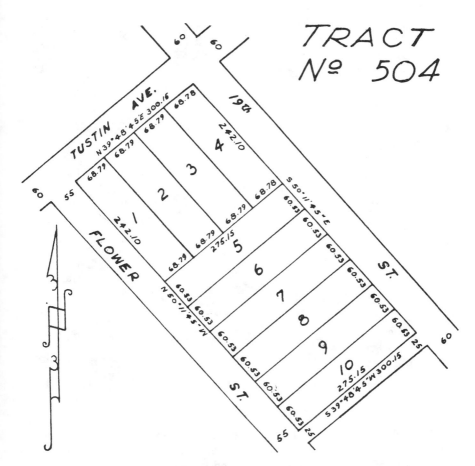

Map for Exercise 12-1.

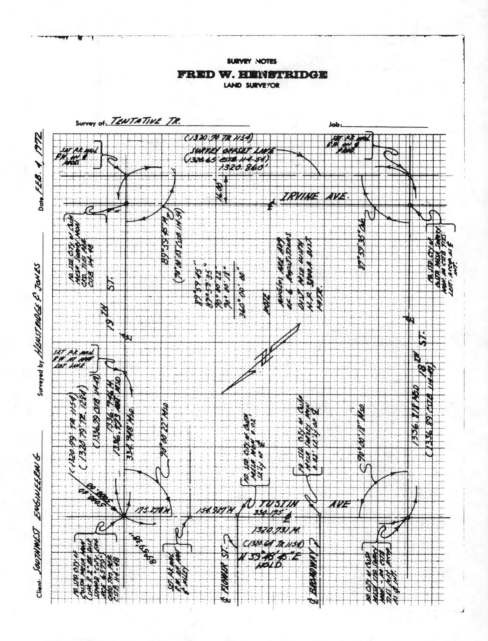

The field notes above show the work done for the boundary control of the new tract while the notes on the next page show field measurements of the record lots to be used in designing the new layout.

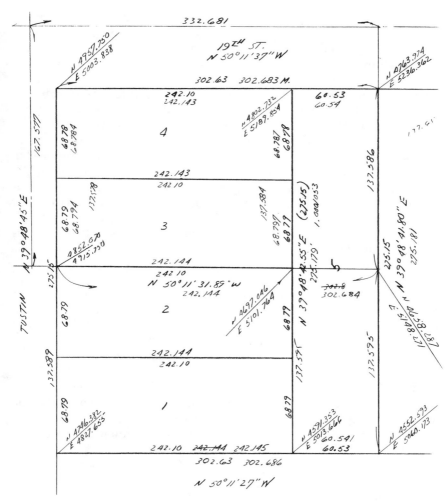

Field Notes for Exercise 12-1.

*   *   *

Condominiums

Insofar as drafting is concerned, this form of development consists of two types of drawings: maps, with or without plans and profiles, and the building layouts tied to the boundary of the property within the map area.

Usually the map is essentially a one-lot tract map with all the units in the one lot. There is another form known as the McKeon Condominium which is composed of multiple lots with a small group of units within each lot. In the first form, the area not covered by buildings and their appurtenances, such as patios, balconies, etc. is devoted to private driveways, utilities, recreation facilities and landscaping.

With the second form, the lots front on a regular street in the tract, and a certain number of those lots are reserved for recreation facilities while landscaping is done around each building on each lot.

The diagrammatic floor plans, showing their inter-relationship to each other and to the property boundary, are drawn on a separate set of sheets; sometimes these are made the same size as the tract map, following with consecutive page numbers, and recorded with the map; alternatively, they may be made the same size or smaller and reduced to the size of the Declaration of Restrictions and recorded with it.

The condominium concept is applicable not only to residential use but also to business and medical offices. On occasion, existing apartment or business buildings are converted into condominium projects in which case it is necessary to survey and draw "as built" plans. If the improvements are on a lot or parcel shown on an already recorded map, there is usally no requirement for making and filing a new map.

A condominium is an estate in real property and consists of an undivided interest in a portion of the property (land) and a separate interest in a certain space, sometimes called "a cube in space," in a building on the property. This "space" interest may be either fee or leasehold and it can be for residential, industrial, medical or commercial use.

The condominium may also include another interest in a different part of the property such as a balcony, patio, garage etc.

The "cube in space" is often furthermore defined as bounded by the interior faces of the walls, the ceiling and the floor; consequently, the diagrammatic plans are intended to reflect these facts.

Items normally included in a Condominium Plan package are: —
  1. Owner's Certificate.
  2. Acknowledgements.
  3. Engineer's Certificate.
  4. Definitions.
  5. General Notes.
  6. Survey of Project.
  7. Building Location Map.
  8. Basis of Datum.
  9. Vertical Elevation Tabulation.
  10. Diagrammatic Floor Plans.
  11. Typical Cross-section.
  12. Parking and Easement Plan.

Figures 12-8 through 14 are copies of selected sheets from such a package for a residential development. In this case there are three lots in the subdivision shown on the Tract map in Figure 12-8.

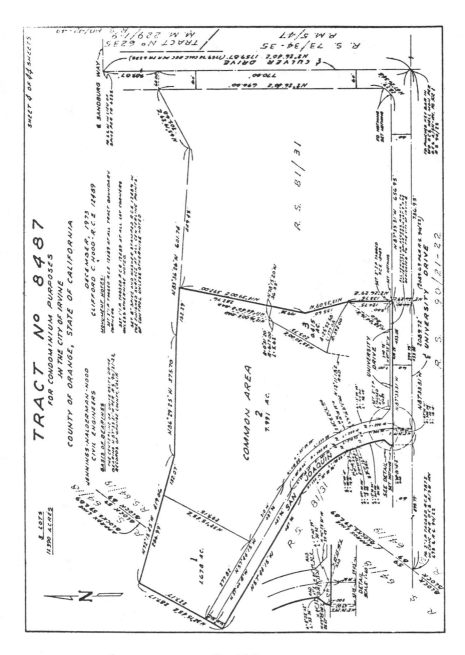

Fig. 12-8

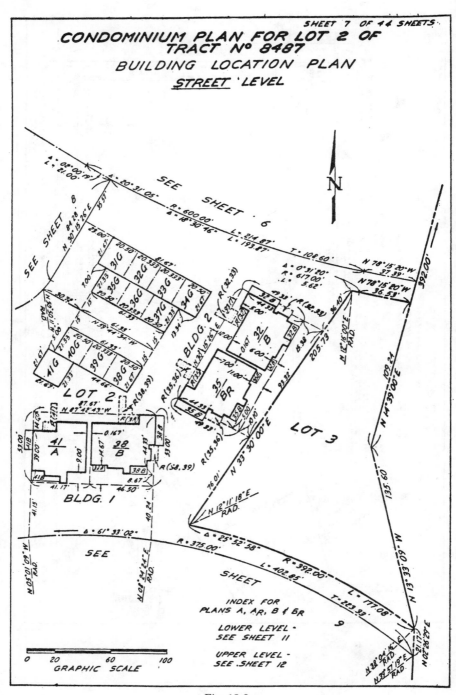

Fig. 12-9

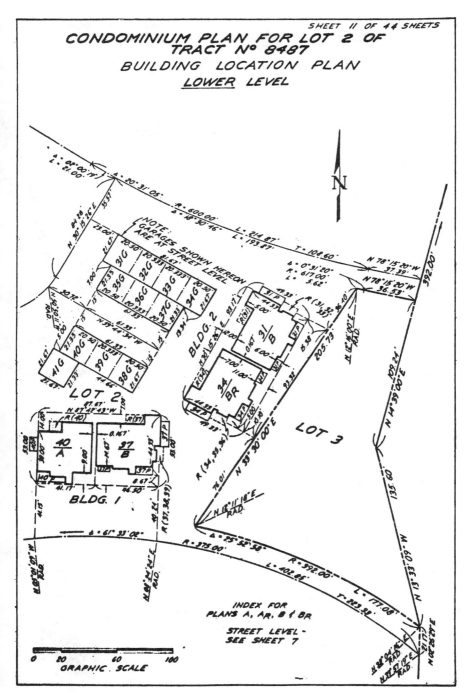

Fig. 12-10

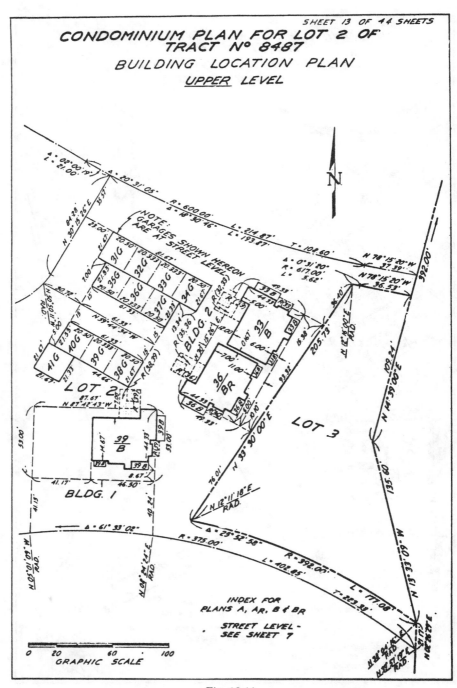

Fig. 12-11

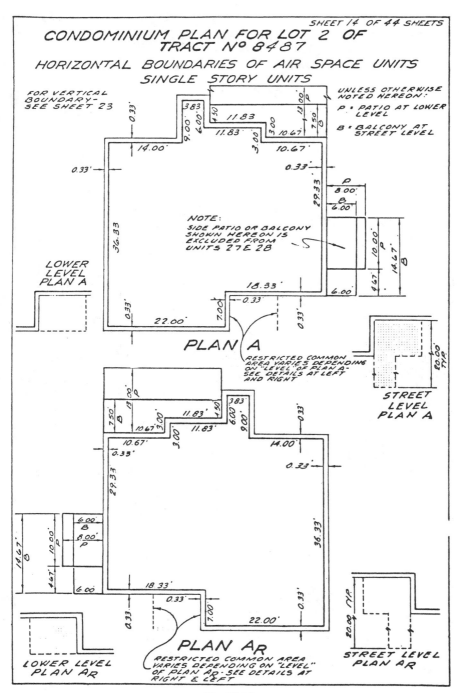

Fig. 12-12

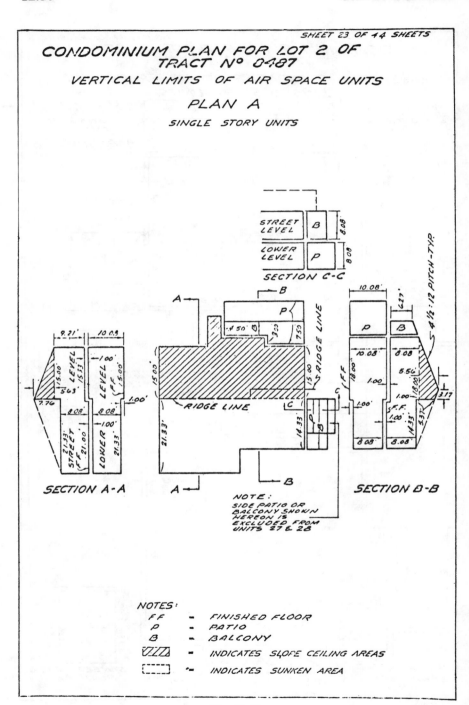

Fig. 12-13

SHEET 43 OF 44 SHEETS

CONDOMINIUM PLAN FOR LOT 2 OF TRACT Nº 8487

TABLE OF FINISHED FLOOR ELEVATION

LEGEND

F F = FINISHED FLOOR
L L = LOWER LEVEL
S L = STREET LEVEL
U L = UPPER LEVEL
R E = RESIDENTIAL ELEMENT
G E = GARAGE ELEMENT

PLAN D

UNIT No	D1 & D1V F F R E	D1 & D1V F F G E	UNIT No	D2 & D2V F F R E	D2 & D2V F F G E	UNIT No	D3 & D3V F F R E	D3 & D3V F F G E	UNIT No	D4 & D4V F F R E	D4 & D4V F F G E	UNIT No	D5 & D5V F F R E	D5 & D5V F F G E
9	75.90	85.00	48	72.40	81.50	7	75.90	85.00	12	73.90	83.00	52	73.90	83.00
15	73.70	82.80				55	73.40	82.50	42	61.40	70.50	59	49.90	59.00
45	62.40	71.50				61	51.90	61.00						

PLAN D - REVERSED

UNIT No	D1R & D1VR F F R E	D1R & D1VR F F G E	UNIT No	D2R & D2VR F F R E	D2R & D2VR F F G E	UNIT No	D3R & D3VR F F R E	D3R & D3VR F F G E	UNIT No	D4R & D4VR F F R E	D4R & D4VR F F G E	UNIT No	D5R & D5VR F F R E	D5R & D5VR F F G E
3	71.90	81.00	NONE			8	75.90	85.00	14	73.90	83.00	11	75.90	85.00
6	73.90	83.00				56	73.40	82.50	44	61.40	70.50			
51	72.40	81.50				60	51.90	61.00						
54	73.90	83.00												
57	49.90	59.00												
62	51.90	61.00												

Fig. 12-14

Three-dimensional Drawings

A perspective drawing is of tremendous help in assisting the comprehension of a difficult three-dimension legal description. Figures and notes on the drawing are easier to perceive than searching for them in an extensive metes and bounds description even though the latter may be required for the legal document.

A typical easement for a bridge is a combination of an elongated cube-like form to encompass the bridge itself plus two or more vertical cylindrical or cubical forms standing between the bridge-in-the-air and the ground for support structures. The perspective drawing in Figure 12-15 illustrates an example of the relationship between the physical structure and the parts of the easement. Following is the description of the parts.

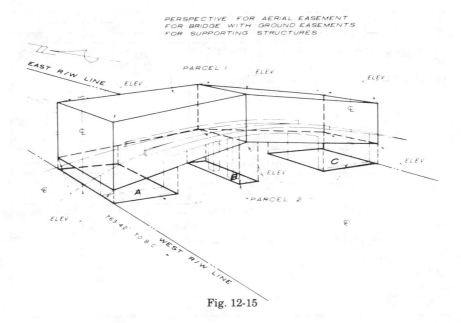

Fig. 12-15

PARCEL 1:

An aerial easement for road and bridge purposes 40.00 feet in width over and above that portion of Section 12, T 4 S, R 8 W of the _____ Meridian in the county of _____, state of _____according to the official plat of said land filed in the District Land Office, the center line of which is described as follows:

> Beginning at a point on the west line of the 300 foot right of way described in deed to the Division of Highways, state of

_____, recorded May 16, 1965 in book
____ page ____ of Official Records, distant North 763.42 feet
from the southerly terminus of that certain course in said
deed cited as "South 1259.62 feet to the beginning of a curve";
thence at right angles East 300.00 feet to the east line of said
right of way; the bottom plane of the vertical space contained
within said easement shall be (based on county surveyor's
B.M. No. 362 Circuit No. 4, checked April 9, 1966, CSTB
140-A page 56) from elevations of 78.52 feet on the west line
of said right of way, to elevations of 90.78 feet on the center
line of said 300 foot right of way to elevations of 79.08 feet on
the east line of said right of way; and the top plane of the
vertical space of said easement shall be 20.00 feet above said
elevations cited for the bottom plane.

In essence, the fundamental descriptions are planimetric and the
elevations which control the vertical limits are set in a complementary
description. Figure 12-16 shows the plan of the horizontal control for the
example bridge easement.

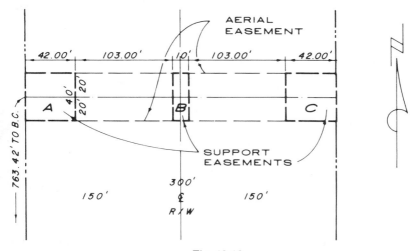

Fig. 12-16

PARCEL 2:
40 foot wide easements for support and maintenance of a bridge
within the aerial easement described in Parcel 1 above, over and
above those portions of Section 12, T 4 S, R 8 W of the _____
Meridian in the county of _____, state of
_____, according to the official plat of said
land filed in the District Land Office, the top plane of said ease-

ments being coincident with the bottom plane described in Parcel 1 above and the center lines of which are described as follows:

EASEMENT A:

Beginning at a point on the west line of the 300 foot right of way described in deed to the Division of Highways, state of _____, recorded May 16, 1965 in book ___ page ___ of Official Records, distant North 763.42 feet from the southerly terminus of that certain course in said deed cited as "South 1259.62 feet to the beginning of a curve"; thence at right angles East 42.00 feet.

EASEMENT B:

Beginning at the above described point on the west line of said 300 foot right of way; thence at right angles East 145.00 feet to the true point of beginning; thence East 10.00 feet.

EASEMENT C:

Beginning at the above described point on the west line of said 300 foot right of way; thence at right angles East 258.00 feet to the true point of beginning; thence East 42.00 feet to the East line of said right of way.

Again the attachment of a map and perspective drawing can be very enlightening. If a perspective is too difficult to obtain, even a plan and vertical cross section would be helpful.

A metes and bounds description around a multi-sided space may be accomplished by the method of a continuous direct perimetrical delineation, but, depending on the shape, it may become extremely involved and unwieldy. To illustrate a simple form, consider the matter of taking a vertical cut out of a street for the purpose of lowering the grade. A map, or a perspective drawing "attached hereto and made a part hereof" is always helpful in clarifying the relationship of the different planes. See Figure 12-17.

That portion of the West 30.00 feet of North Nicolas Avenue adjoining Lot 12 in Tract No. 201 in the City of _____ _____, County of _____, State of _____, as per map recorded in Book 13, Page 6, Miscellaneous Maps in the office of the County Recorder of said county, bounded northerly by the easterly prolongation of the northerly line of said Lot 12 and bounded southerly by the easterly prolongation of the southerly line of said Lot 12 and included within the vertical section lying above the following described plane:

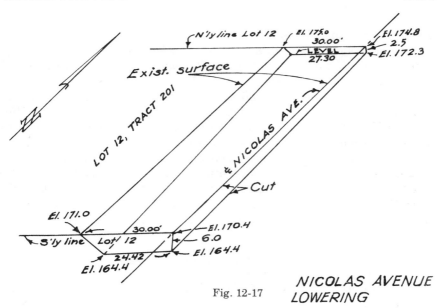

Fig. 12-17

NICOLAS AVENUE
LOWERING

Beginning at the northeast corner of above described West 30.00 feet, the existing surface elevation of said point being Elevation 174.8 feet per city datum as shown on Plan File No. 1264 on file in the office of the City Engineer of said City; thence vertically downward 2.5 feet to Elevation 172.3, said point being the True Point of Beginning; thence, along the east line of said West 30.00 feet, Southerly and descending to Elevation 164.4 in the southeast corner of said west 30.00 feet; thence, along the southerly line of said West 30.00 feet, Westerly 24.4 feet to a point at Elevation 164.4 in a line having a downward slope of 1 foot vertical to 1 foot horizontal, measured from the existing surface elevation (Elevation 171.0) at the southeast corner of said Lot 12; thence, Northerly and ascending to a point at Elevation 172.3 in the northerly line of aforesaid West 30.00 feet, said point being in a line having a downward slope of 1 foot vertical to 1 foot horizontal, measured from the existing surface elevation (Elevation 175.0) at the northeast corner of said Lot 12; thence, along the north line of said West 30.00 feet, Easterly 27.3 feet to the True Point of Beginning.

TOGETHER WITH a longitudinal section of triangular cross section, the lower plane of which is that certain plane connecting the 1 foot vertical to 1 foot horizontal line in the southerly line of said West 30.00 feet with the 1 foot vertical to 1 foot horizontal line in the northerly line of said West 30.00 feet, as said lines are hereinabove described.

Agreement-Line Maps

This kind of map may be developed into a deed type or a recorded map. The objective is the same but the *modus operandi* of the attorney will determine the form to use.

The features to be shown on such a map are essentially the line to which the parties agree, its relationship to all lines of *title* ownership concerned with its establishment, any physical conditions pertinent to the agreed-upon line, references to matters of record and clear outlines with ties of the parcels which each party is to convey to the other in order to effect the final status of ownership on each side of the new line. See Figure 12-18. The parcels will either be described by metes and bounds, or *IF* the map carries sufficient detail, reference to parcels shown on the "map attached" or "map recorded in Book ＿＿ at page ＿＿" will suffice.

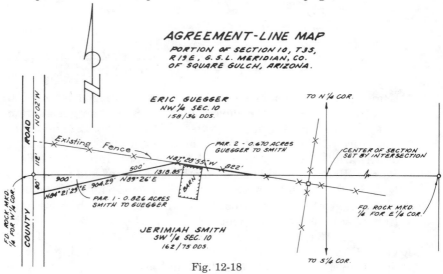

Fig. 12-18

Road Realignment

Sometimes roadways are realigned and maps must be made to show the relationship between the old and the new positions. This is also true with railroads and sometimes pipe lines.

If the new alignment is office designed and calculated, the map is drawn prior to construction. If survey information of the field location of the realignment is furnished, calculations are in order to check the closure of the new position against the old and make any mathematical adjustments before creating the map.

The field notes in Figures 1-23 and 24 show office calculations given to the survey crew for setting the control in the field.

A part of a more extensive change of alignment is shown in Figure 12-19 where many curves are required because of the hilly terrain.

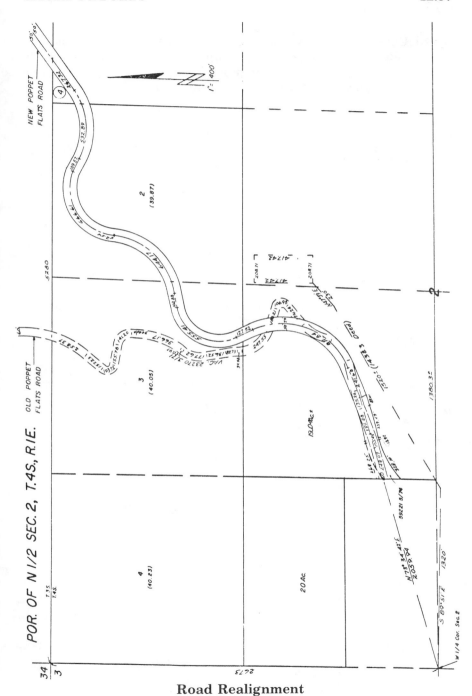

Road Realignment

Fig. 12-19

EXERCISE 12-2: Copy the drawing of the "Existing Roadway" shown in Figure 12-20, complete the calculations for the curves, then add the "Proposed Realignment of 60' Roadway" in Figure 12-21 and complete all of its calculations.

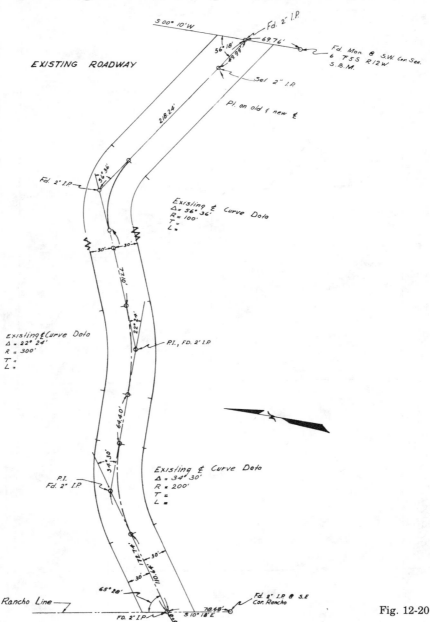

Fig. 12-20

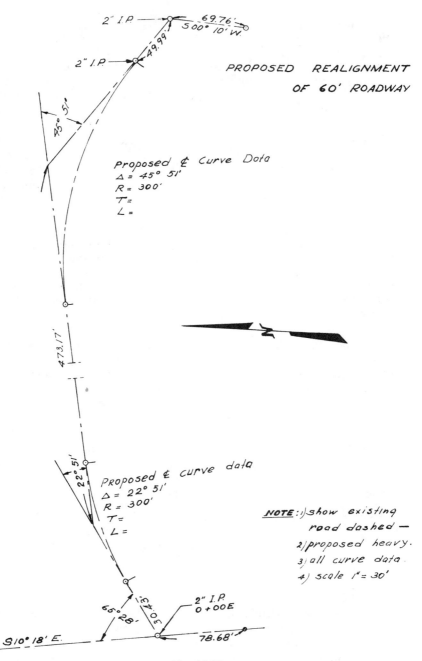

Fig. 12-21

* * *

Leasehold Maps

There are several areas in which leases of property, or leases for improvements on property, are made for the exercise of certain rights. Shopping centers may be subdivided into parcels and leases set on certain parcels or the whole area may be kept as a single unit or one-lot subdivision and the leases spotted here and there. In either case, your map will show the outlines with adequate ties to the perimeter so that each lease area can be properly surveyed and marked on the ground.

The same principle can be applied to an industrial complex with one added possibility of railroad service, in which case the right-of-way will need to be located in cooperation with the railroad company's engineer.

Again these maps may be small documentary size to accompany the leasehold papers or large enough to file in the office of the public recorder.

In the matter of oil and gas leases, however, you have two general types, single-property leases and community leases; the first refers to a lease covering only one ownership (which considers joint tenants and/or co-owners as comprising one ownership), while the second refers to a lease covering a *group* of separate parcels of land in separate owner-ships. Another factor in the community lease is that the oil and/or gas is removed through one or more properties for the benefit of all in that lease.

The maps for such leases are in fact property ownership maps but, for the benefit of the lessee, carry an additional reference to the lease or to a code designation of the lease. In addition, there must be reference to the interest of those owners of rights in and to the petroleum, crude oil and natural gas who are not the owners of the fee interest in the land.

The community lease map may show all this information plus a heavy outline of the limits of operation, or show only the outline of the commu-nity lease area and keep the other information on separate sheets.

One other item to show on such maps is easements; they may be needed to get to the well site for its development and/or for pipe lines for transportation of the well's production.

Two other maps that are for special purposes relate each to the same kind of data but one picturizes a legal description which is yet to be recorded while the other illustrates information which is already a matter of record.

Exhibit Map

On occasion an Exhibit Map is requested, or needed, to facilitate the locating of an easement, a leasehold, an isolated parcel, etc. which may or may not reach the status of statutory recordation. Figure 12-22 shows one specimen of this. On the other hand, an Exhibit Map made for the specific purpose of being recorded with the document which it illustrates is often referred to as a "Deed Map." See Figure 12-24.

Deed Maps

As the name implies, this type of map is attached to a deed and usually by recitation, it is "made a part hereof." The size is usually the same as legal documents accepted by the office of public records.

It may be less or more detailed than other maps which are ultimately recorded, depending on the circumstances and therefore the requirements involved. The "deed" map is not checked by a public official so you must be more careful and double check before releasing it.

In all situations, your "deed" map should have excellent ties and references shown on it and be so complete in detail that there can be no dispute as to the correct establishment of the property lines involved.

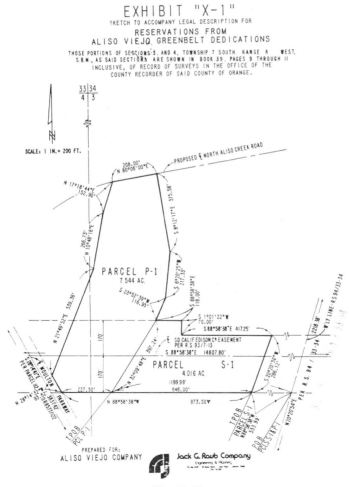

Fig. 12-22

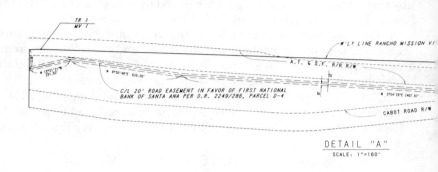

DETAIL "A"
SCALE: 1"=160'

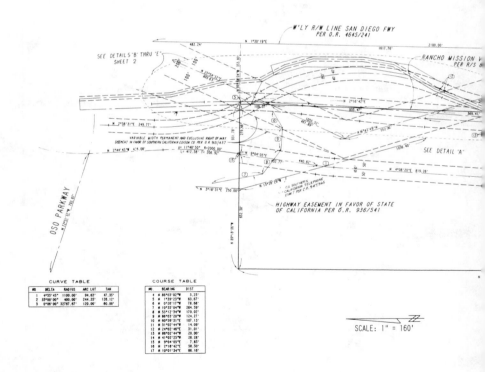

CURVE TABLE			
NO	DELTA	RADIUS	ARC LGT	TAN
1	4°55'45"	1100.00'	94.63'	47.33'
2	35°00'00"	400.00'	244.35'	126.13'
3	0°06'00"	32767.67'	120.00'	60.00'

COURSE TABLE		
NO	BEARING	DIST
4	N 86°03'02"W	3.25'
5	N 1°39'23"W	63.67'
6	N 0°26'17"W	76.66'
7	N 10°35'04"W	284.59'
8	N 53°12'34"W	170.02'
9	N 86°03'20"W	124.27'
10	N 60°59'31"E	107.13'
11	N 31°02'44"W	14.09'
12	N 24°02'46"E	31.01'
13	N 86°02'44"W	20.00'
14	N 41°02'25"W	28.26'
15	N 9°04'05"E	7.85'
16	N 2°18'42"E	56.50'
17	N 10°01'34"E	66.16'

SCALE: 1" = 160'

Compiled Record Data Map

This type of map shows the detailed juxtaposition of facts of record. The complexity of it depends upon two things: the number of items to be shown and the amount of detail in each legal description. The example shown in Figure 12-23 involves eleven documents.

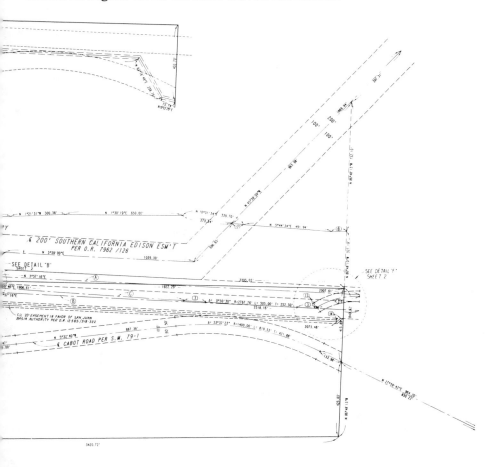

COMPILED RECORD DATA MAP
PRELIMINARY - SUBJECT TO
TITLE REPORT ANALYSIS & BDRY. SURVEY
"GALIVAN" AREA - P.A. 72

Ⓐ C/L "OLD HIGHWAY 101" IN FAVOR OF
    STATE OF CALIFORNIA PER O.R. 149/57
Ⓑ C/L 100' SOUTHERN CALIFORNIA RAILROAD
    R/W (A.T. & S.F.) PER DEED 39/277
Ⓒ C/L 100' RAILROAD RIGHT OF WAY PER A.T. &
    S.F. R/W PLAT 240-20127 (O.R. 1121/499)

Compiled Record Data Map

Fig. 12-23

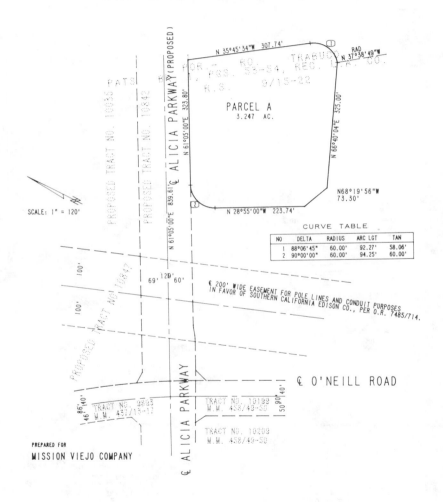

C H I Q U I T A   S U B S T A T I O N

A PORTION OF THE RANCHO TRABUCO IN THE UNINCORPORATED TERRITORY OF THE
COUNTY OF ORANGE, STATE OF CALIFORNIA, AS SHOWN ON THE MAP RECORDED DECEMBER
19, 1867, IN BOOK 1, PAGES 53 AND 54 OF PATENTS, IN THE OFFICE OF COUNTY
RECORDER OF LOS ANGELES COUNTY, STATE OF CALIFORNIA.

PARCEL A
3.247 AC.

CURVE TABLE

NO	DELTA	RADIUS	ARC LGT	TAN
1	88°06'45"	60.00'	92.27'	58.06'
2	90°00'00"	60.00'	94.25'	60.00'

SCALE: 1" = 120'

℄ 200' WIDE EASEMENT FOR POLE LINES AND CONDUIT PURPOSES
IN FAVOR OF SOUTHERN CALIFORNIA EDISON CO., PER O.R. 7485/714.

℄ O'NEILL ROAD

PREPARED FOR
MISSION VIEJO COMPANY

Sketch to Accompany Legal Description

Fig. 12-24

R/W or Strip Map

Right-of-way maps are generally produced under one of two classifications: 1, showing ownerships and outlines of parcels for the proposed acquisition of land based either on a preliminary survey line (called a "P" line) or developed from orthophoto overlay maps; or 2, the final alignment with relationship of construction to R/W and property lines **shown in plan and with the profile to match it. Such a plan is shown in Figure 12-6.**

In the case of a parkway or freeway interchange involving several acres of land, the map for the right-of-way layout and acquisition will probably be one large sheet or possibly four medium sized sheets, one for each quarter of the development. Obversely, for the continuous run of straight stretches of roadway, the right-of-way layout and acquisition information will be shown on narrow strip maps.

In both cases, the plan-and-profile drawings are done on either the standard size sheets or a continuous roll.

The ownership maps should be carefully prepared *from the legal descriptions of each parcel*. In this way, any problems relating to GAPS or OVERLAPS of adjoining parcels will be shown so the land acquisition officer in charge of operations can see the problem and alert his attorney or title company to get more information about it.

In addition, this will help clarify and consequently assist the person who must write the new legal descriptions for the parcels to be acquired. (Review a typical problem discussed in Chapter 3 under "Rights-of-Ways.")

The parcels to be taken are given numbers for subsequent identification and tabulation. If condemnation is necessary, the descriptions with the parcel numbers and map are included in the proceedings.

When the acquisition lines of the R/W boundary are established, the plans and profiles for construction are developed as previously discussed.

A final R/W alignment and location in relation to lot lines is shown in Figure 12-25.

A.L.T.A. Survey Maps

This particular type of map and survey is governed by certain requirements set forth by the American Land Title Association for use related to the issuance of Extended Coverage Policies of Title Insurance. It is extensively detailed, not only in regards to all pertinent survey information, but also as to the relationship of physical things to the surveyed lines of ownership for both fee title and easement holders. Because of this you need to work very closely with the title company that will issue the Policy on the subject land.

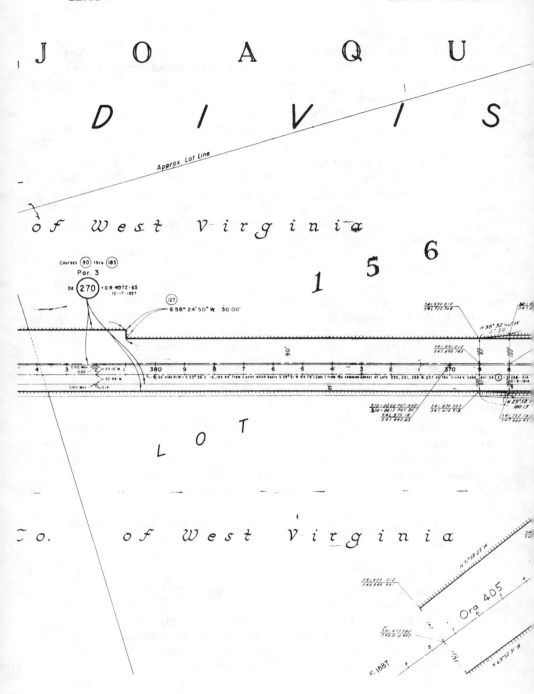

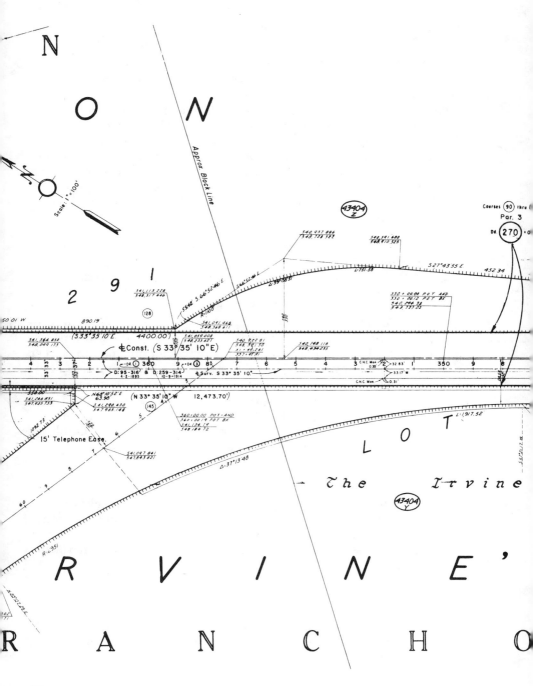

Fig. 12-25

Preparation for A.L.T.A. Work

First of all, verify the description of the property involved and analyze it enough to know if there are any inherent problems and, if so, determine what remedial action will be necessary. Some of this may not be apparent or determinable until after the survey is made.

Obtain copies of all documents to which the preliminary title report refers because you need to plot the easements, rights-of-ways and boundary lines pertinent to your job. Getting copies of all maps applicable to the property is likewise a tremendous help. By doing this *before* the field work is done and giving it to the surveyor to take with him, he has on the one hand facts in question which can be considered in the field, and on the other hand, assistance from additional information by which to accomplish the survey.

Map Details

It is of the utmost importance that there be sufficient evidence shown on the final map to prove the establishment of all boundaries; from an examination of the map submitted by the surveyor, his adoption of the location delineated thereon should be clearly apparent. This often requires a showing of actual measured distances and bearings around the entire block; city or county field book references for all adjoining street center lines; and any monuments found which indicate property lines. Where occupation is chosen as the best location, enough ties to buildings, walls, fences, surveyor's monuments, etc., must be shown to indicate the pattern of occupation within the block and the actual or estimated age, number of stories and structural materials of buildings influencing such choice indicated, particularly those on subject and adjoining lands.

Maps should also show:

1. Bearings and distances on exterior boundaries of parcel surveyed. If said parcel is composed of all or portions of several lots or other legal subdivisions, the boundaries of each should be indicated by dotted lines and the proper lot number or legal subdivision designation shown.
2. Location and dimensions of
   (a) buildings or other structures on the surveyed parcel and basement extensions, if any, beyond property lines, and
   (b) any buildings or other structures erected on adjoining lands within five feet of the surveyed parcel. If adjoining land is vacant or used for other than building purposes, so state.
3. All party walls, the thickness of the portions thereof on each side of the property line; and the nature of the use of said walls on each

side such as "integral wall of both buildings," "wall of adjoining building used as support for roof of this building," etc.

4. Whether walls are plumb and if they extend beyond property lines. If a building or other structure is exactly on the property line, so state. If the building is over the line, show the amount by figures. Do not rely on drafting.

5. Extension of cornices, eaves, window ledges, pipes, canopies, base blocks, conduits or other projections over property lines in either direction with dimensions.

6. All fences and walls (including gates, openings, doorways, passageways, etc.), driveways and other improvements along property lines and dimensions of same. If determinable, show ownership of boundary fences and walls. Community driveways should be so shown.

7. All manholes, conduits, drains, pipe lines (including abandoned lines), railroad tracks, roadways, pathways and similar matters which may indicate use by other parties and, where possible, indicate property served.

8. All wires and cables (including their use) crossing, entering or leaving the parcel surveyed, except the ordinary two or three-wire service drops to said parcel, which may be omitted; all wire bearing poles on or within ten feet of said parcel and pole numbers; amount of cross arm or wire overhang affecting said parcel; and all anchor and guy wires affecting said parcel.

9. The location of any topographic features such as natural or artificial water courses.

10. Curbs and walks on or adjoining the parcel surveyed with dimensions from property line, and position of any sidewalk markers indicating claimed location of property line.

11. Dotted lines outlining areas affected by record easements with deed references, if available.

12. All signs and billboards.

13. House numbers as they exist. If non-existent, show assigned or recognized number, if available, and source of information.

14. Caption and title sufficient to identify the parcel surveyed including lots, blocks, tract, map reference, city (if any), county and state.

15. North point and scale.

16. A certificate by the surveyor substantially as follows:

"I hereby state that this survey was made for the purpose of an application for title insurance; at the request of _(name of client)_ under my supervision on ___(date)___ and that said survey correctly shows the relation of buildings and other

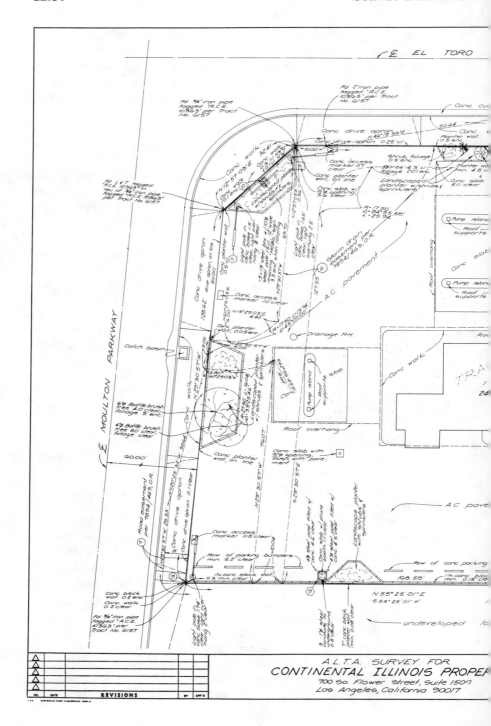

A.L.T.A. SURVEY FOR
CONTINENTAL ILLINOIS PROPER
700 So. Flower Street, Suite 1507
Los Angeles, California 90017

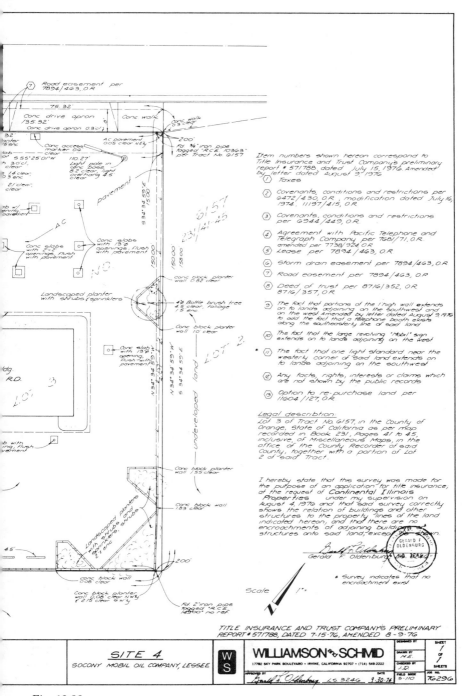

Fig. 12-26

structures to the property lines of the land indicated hereon; and that there are no encroachments of adjoining buildings or structures onto said land, except as shown."

(signed) _____

Applicable Registration No. (LS or RE):

CERTAIN OF SAID REQUIREMENTS MAY BE RELAXED UNDER SPECIAL CIRCUMSTANCES.

Examples:

(a) The interior portion of the surveyed parcel is improved by various structures or by structures of complicated design, the delineation of which would add materially to the survey cost.

(b) The surveyed parcel is unrestricted or there are no buildings thereon less than five years old; there are no improvements within five feet of any of the boundary lines of said parcel; and the surveyed measurements and location of said parcel are substantially the same as the dimensions and location thereof shown on the recorded map of the subdivision.

In such cases, and in cases where a full survey is not required, consult with the title company as to procedure.

*

In reviewing other survey maps, be sure to analyze the method used by that surveyor before accepting it because the courts have held to "facts that a correct survey would show." This also leads to the discussion of "accuracy vs. correctness." You might have a traverse closure of 1:40,000 but if the measuring tape was off 0.02'/100', your result would be incorrect.

Delineation of the differences in position of any monuments from that of the lines and corners of record title is essential.

If you have a railroad for boundary, be certain whether its right-of-way is controlled by the physical location of the track *or* a line actually described in a deed.

Show overhang of trees, both onto and off of subject property.

Also show *obvious* easements, if any, such as auto trails or cattle paths across the subject property and label them.

Wells of any type with pipe lines to and from them warrant valuable consideration.

Figure 12-26 shows a completed ALTA survey map with much of the detailed information heretofore discussed.

EXERCISE 12-3: From the following legal description of the property and the descriptions of the easements, draw an ALTA map showing all those matters plus the details reported in the field notes by the survey crew.

Description: Lot 7 of Tract No. 8590 in the City of Tustin, County of Orange, State of California, as per map recorded in Book 346, Page 19, of Miscellaneous Maps, records of said County.

Easements:

1. (hereinafter referred to as "Grantor(s)", hereby, grant(s) to SOUTHERN CALIFORNIA EDISON COMPANY, a corporation, its successors and assigns (hereinafter referred to as "Grantee"), an easement and right of way to construct, use, maintain, operate, alter, add to, repair, replace, reconstruct, inspect and remove at any time and from time to time underground electrical supply systems and communication systems (hereinafter referred to as "systems"), consisting of wires, underground conduits, cables, vaults, manholes, handholes, and including above-ground enclosures, markers and concrete pads and other appurtenant fixtures and equipment necessary or useful for distributing electrical energy and for transmitting intelligence by electrical means, in, on, over, under, across and along that certain real property in the County of Orange, State of California, described as follows:

    A strip of land 8 feet in width, lying within said Lot 7, the centerline being described as follows:

    Commencing at the most easterly corner of said Lot 7; thence southwesterly, along the southeasterly line of said lot, a distance of 13 feet to the TRUE POINT OF BEGINNING: thence northwesterly, parallel with the northeasterly line of said lot, a distance of 23 feet.

    Also a strip of land, 6 feet in width, the centerline being described as follows:

    Beginning at a point in the northeasterly line of said lot, distant northwesterly 20 feet from the most easterly corner of said lot; thence southwesterly, parallel with the southeasterly line of said lot, 10 feet, more or less, to the northeasterly line of the above-described 8-foot strip.

2. to Breton Development Co.:

    A non-exclusive easement for vehicular ingress, egress and pedestrian walkway purposes over the southerly 12 feet of said Lot 7, which easement shall be appurtenant to and pass with the fee interest in Lot 8 of said Tract 8590.

3. Those easements shown and dedicated on the map of Tract No. 8590.

TRACT Nº 8590

THE CITY OF TUSTIN, COUNTY OF ORANGE, CALIFORNIA.

GERALD F. OLDENBURG, LS 3246

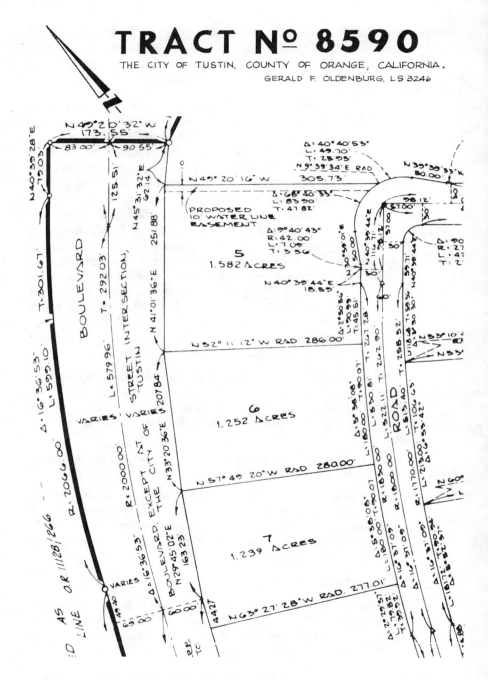

Map for Exercise 12-3

ALTA   78318   2   12
M. Gillen, Barnard, Cassidy   8/25/78
Property Line "A"

-0+14.5   0.3 Rt. to Pnt. where flow line of 3' wide conc. gutt. from N.E. int. w/B.W.

-0+04.5   0.3 Rt. to Bk. of Dr. Apr. @ Pnt where C.F. from No. int. This Pnt. is also the S.E. cor. of A.C. Dr.-0" CF @ this pnt. 6" 2' to No.

-0+00.3   0.3 Rt. to N.E. cor. conc. Dr. Apr.-8.1 Rt. to C.F. @ top of 4' "X" of drive to West.

0+00.8   5.0 Rt. to No. edge of 1' Dia. (@ Base) Fire Hyd.

0+01.6   0.3 Rt. to N.W. cor of conc. Dr. Apr.-8.1 Rt. to C.F. @ top of 4' "X" - West end of 29' wide drive (top to top) to East.

0+04.8   0.3 Rt. to Bk. of Dr. Apr. @ pnt where C.F. from No. int-0" at this pnt.-6" F.C. to No. This pnt. is also the S.W. cor of A.C. Dr.

0+30.2   0.5 Rt. to Bk. of Drive Apr. @ pnt. where C.F. from No. int. 0" C.F. @ this pnt. 6" C.F. 2' to No. This pnt is also the S.E. cor. of A.C. Dr.

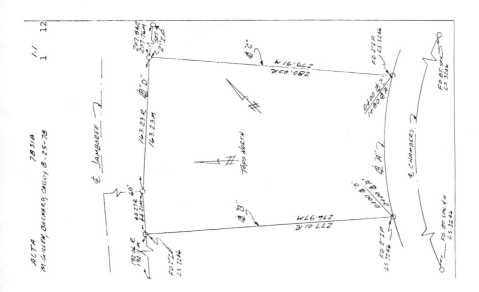

Field Notes for Exercise 12-3

ALTA   78318   4   12

M. Gillen, Barnard, Cassidy 8/25/78

1+68.4   6.0 Rt. to N.W. cor. 1.3 No. & So x 2.0 E & W conc. cable T.V. vault-flush w/t.c.

1+68.5   5.7 Lft. to 2" dia. tree-12' ± tall-5' fol. rad.

1+74.8   2.5 Rt. to N.W. cor. 4.4 No. & So. x 7.4 E & W Steel vault - no identification.

1+82.2   2.5 Rt. to N.E. cor. 4.4 No. & So. x 7.4 E & W Steel vault - no identification.

Base Line "B"

0+04.2   0.5 Rt. to 2" dia. tree-12' ± tall - 2' fol. rad.

0+05.3   4.1 Rt. to Bk. of curb @ ECR of 4' rad. curve (C.F. rad)

0+05.3   3.8 Lft. to Bk. of curb @ BCR of 4' rad. curve (C.F. rad)

0+09.0   0.6 Rt. to BCR of 4'± rad. curve (C.F. rad) ± is to bk of curb.

0+09.0   0.3 Lft. to ECR of 4'± rad. curve (C.F. rad)± is to bk of curb.

0+34.2   14' Rt. to West end of curb @ So face-this pnt. is also the N.W. cor of conc. walk way 8.2' wide No. & So. from this pnt.The walk ⚡'s to S.E.

ALTA   78318   3   12

M. Gillen, Barnard, Cassidy 8/25/78

0+30.7   0.5 Rt. to N.E. cor. conc. Dr. Apr.-8' Rt. to C.F. @ top of 4' "X" E. side of 29' Drive-top to top. Note: Row of sprink runs along bk. of curb-No. side Chambers.

0+40.1   5.5 Rt. to N.W. cor. 1.9 No. & So. x 1.3 E & W conc. Wat met vault-flush w/gr.

0+43.5   0.3 Lft. to So. edge of 7" dia. 4' tall iron fire cont. valve.

0+43.5   3.1 Lft. to So. edge of 4" dia. iron fire cont. valve-3' tall. top ⚡'s,1' to S.W.

0+45.2   4.5 Rt. to N.W. cor. 2.8 No. & So x 1.8 E & W conc. wat met vault-flush w/grass.

0+63.7   5.7 Rt. to No. edge - 9" dia. stone lite pole #206141OE set in 2.5 No. & So. x 1.8 E & W conc. base flush w/t.c.

0+76.3   9.4 Lft. to 4" dia. tree-15' ± tall, 5' fol. rad.

0+82.8   3.6 Lft. to 1" dia. tree-12' ± tall,2' fol. rad.

1+01.5   5.1 Lft. to 4" dia. tree-15' ± tall,6' fol. rad.

1+39.7   4.7 Lft. to 3" dia. tree-15' ± tall,3' fol. rad.

Field Notes for Exercise 12-3

ALTA   78318   6   12
M. Gillen, Barnard, Cassidy 8/25/78

1+23.5 12.1 Lft. to C.F. @ BCR of 4' rad. curve-curves 90° to S.W. & ends @ Bld.

1+44.0 18.8 Rt. to C.F. @ W. edge of 10' wide (No. & So.) planter island, has 2'± rad. @ N.W. & S.W. cors.

1+77.6 20.7 Rt. to C.F. @ W. end of planter island. This pnt. is the ECR of a 2' rad. curve of curb to East. 90°± Delta & the BCR of 2' rad. curve of curb,running to N.E. @ 45°±.

1+78.6 18.7 Rt. to Angle Pt. flow line of 3' wide conc. gutt.

1+86.6 12.6 Lft. to E'ly. C.F. @ ECR of nose of planter island to West. Delta=180°-Planter is 4' wide.

2+13.3 Base Line crosses nose of 9' wide E & W planter island.

2+15.6 0.1 Rt. to centerline 1' dia. (@ base) Fire Hyd.

2+16   4.9 Rt. to W. edge of 7" dia. steel wat. valve flush w/A.C.

2+17.8 3.9 Rt. to Bk. of curb @ BCR of 4.5' rad. curve (C.F. rad).

2+18.0 4.0 Lft. to Bk. of curb @ ECR of 4.5' rad. curve (C.F. rad).

ALTA   78318   5   12
M. Gillen, Barnard, Cassidy 8/25/78

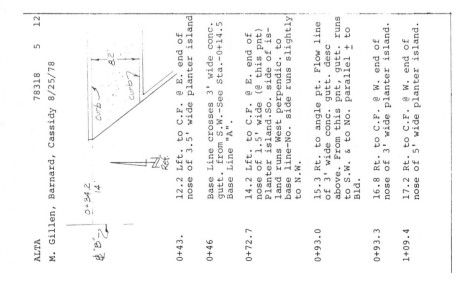

0+43.   12.2 Lft. to C.F. @ E. end of nose of 3.5' wide planter island

0+46   Base Line crosses 3' wide conc. gutt. from S.W.-See Sta.-0+14.5 Base Line "A".

0+72.7 14.2 Lft. to C.F. @ E. end of nose of 1.5' wide (@ this pnt) Planter island.So. side of island runs West perpendic. to base line-No. side runs slightly to N.W.

0+93.0 15.3 Rt. to angle pt. Flow line of 3' wide conc. gutt. desc above. From this pnt, gutt. runs to S.W. & to No. parallel ± to Bld.

0+93.3 16.8 Rt. to C.F. @ W. end of nose of 3' wide planter island.

1+09.4 17.2 Rt. to C.F. @ W. end of nose of 5' wide planter island.

Field Notes for Exercise 12-3

ALTA   78318   8   12
M. Gillen, Barnard, Cassidy 8/25/78

*2+30   Base Line crosses toe of 7'± tall earth berm.

2+48   Base Line crosses top 7'± tall earth berm.

2+54   Base Line crosses top 7'± tall earth berm.

2+71   Base Line crosses toe 7'± tall earth berm.

2+77   Check in.

Base Line "C"

-0+02.3 0.1 Lft. to E. edge of 2" x 4" redwood header @ beg. to No.-row of sprink runs along W. side of header.

0+12.4   10.7 Lft. to S.E. cor of 7.8 No. & So. x 5.9 E & W conc. slab flush w/grass-slab has 5.6 E & W x 46 No. & So. x 5' tall metal box on S'ly 1/2.

0+24   0.9 Lft. to E. edge of 12" dia. plastic sprink. control vault flush w/grass.

0+26.7   8.6 Lft. to 5" dia. tree 15'± tall-4' Fol. Rad.

0+34.2   10.30 Lft. to S.E. cor. of 20'± tall conc. tilt-up Bld. See Detail Page 12.

ALTA   78318   7   12
M. Gillen, Barnard, Cassidy 8/25/78

2+26.0   3.8 Lft. to Bk. of curb @ No. end of curb-@ this pnt. a .5' tall 8" thick conc. block wall begs. to No. 3.4' to No. of this pnt, wall begins to step up, 5 steps, each .7 up & .7 out.

2+30.5   4.3 Lft. to W. face of 8" thick conc. block wall @ pnt. where C.F. from West int.-this pnt. is also the N.E. cor. of A.C. parking area.

*2+29.7 4.4 Rt. to angle pt. bk. of curb Angles to So. & to N.E.

2+29.7   3.3 Lft. to .8' tall 3/4" dia. sprink.

2+30.6   6.6 Rt. to Pt. where bk. of curb from S.W. (desc. above) ends @ West end (No. face) of 8" thick 3' tall conc. block wall - A 1.2' tall, 2" thick wood ret. wall runs along back of curb btwn this pnt. & angle pt. desc. @ Sta. 2+29.7.

2+33.3   3.5 Lft. to angle pt. 8" thick block wall-out is to E. face-from this pnt. wall angles to So. & to West-@ this pnt. wall is 3.5' tall So. side-1' tall No. side.

2+41.   6.4 Rt. to 2" dia. pine tree 8'± tall 3' Fol. Rad.

Field Notes for Exercise 12-3

ALTA   78318   10   12
M. Gillen, Barnard, Cassidy 8/25/78

1+87.4 — 17.70 Lft. to N.E. cor. of 20'+- tall conc. tilt-up Bld. See Detail Page 5. This pnt. is also the West end of curb from East.

1+87.7 — 5.2 Lft. to angle pt. bk. of curb desc. above-angles to No. & to West-2" x 4" redwood header runs from this pnt. East & ends @ West face of 2 x 4 header running No. & So.-curb to No. has row of sprink running along E. side.

1+95.2 — 3.1 Lft. to 1" dia. tree 12'+- tall-3' Fol. Rad.

2+08 — 3.3 Lft. to 1" dia. Pine 6'+- tall-2' Fol. Rad.

2+18.4 — 3.6 Lft. to 1" dia. Pine 5'+- tall-3' Fol. Rad.

2+23.1 — 6.7 Lft. to Bk. of curb @ No. end-So. end of 8" thick 3.5' tall conc. block wall (E. Face)

2+28 — 1.8 Lft. to East edge @ No. end of 2" x 4" redwood header.

2+29 — 4.2 Lft. to 1" dia. tree 12'+- tall-2' Fol. Rad.

2+29.0 — Base Line crosses toe of 9'+- tall earth berm.

ALTA   78318   9   12
M. Gillen, Barnard, Cassidy 8/25/78

0+37.1 — 9.5 Lft. to 1" dia. x .8 tall gas pipe.

0+44 — 5.6 Lft. to 1" dia. tree 12'+- tall-2' Fol. Rad.

0+57 — 6.0 Lft. to 5" dia. tree 12'+- tall-4' Fol. Rad.

0+77 — 6.2 Lft. to 2" dia. tree 12'+- tall-2' Fol. Rad.

0+93.7 — 9.1 Lft. to 2" dia. tree 12'+- tall-3' Fol. Rad.

1+00 — .5 Lft. to E. edge of 2" x 4" redwood header.

1+10.5 — 5.8 Lft. to 2" dia. tree 10'+- tall-3' Fol. Rad.

1+29.2 — 6.1 Lft. to 2" dia. tree 12'+- tall-4' Fol. Rad.

1+46.2 — 10' Lft. to 2" dia. tree 12'+- tall-5' Fol. Rad.

1+62.1 — 6.8 Lft. to 2" dia. tree 12'+- tall-2' Fol. Rad.

1+86.0 — 17.1 Lft. to 1" dia. gas pipe @ pnt. of exit from ground-runs 2' to No. (Flush w/ground) & ends.

Field Notes for Exercise 12-3

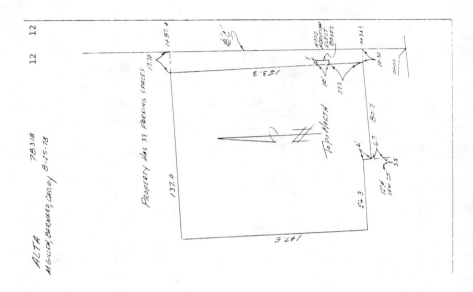

ALTA   78318   11   12

M. Gillen, Barnard, Cassidy 8/25/78

2+33.7   6.9 Lft. to angle pt. conc.
          block wall—angles to So. & to
          West—3.8' tall So. side—0' tall
          No. side.

2+39.6   7.6 Lft. to 2" dia. tree 12'+
          tall—3' Fol. Rad.

2+52   Base Line crosses top of 9' tall
          berm.

2+58   Base Line crosses top of 9' tall
          berm.

2+80   Base Line crosses toe of 9' tall
          berm.

          Base Line "D"

Toe of 9'+ tall berm runs + on property
line – No other improvements.

Field Notes for Exercise 12-3

* * *

Court Case Exhibits

This particular field of map making leans toward the dramatic expression because the area or the special subject needs to be magnified enough to impress its importance on the court.

One map may not be enough; it may be necessary to "zero in" on the target with a very much enlarged view of the subject.

In any event, controlling street names, highways, railroads, cross streets, etc., should appear in large lettering that is much larger than the surrounding information so that anyone 15 to 25 feet away can read it. After you draw the illustration, set the board 25 feet from you and see if you can read it. To further help direct the eyes of the viewer, a large arrow gives good assistance as shown in Figure 12-27.

Fig. 12-27

The "x marks the spot," a house, two cars bumped together, a school crossing, or whatever the attention is to be focused upon, should be set forth in contrasting colors and again *large* enough to be easily seen. See Figure 12-28.

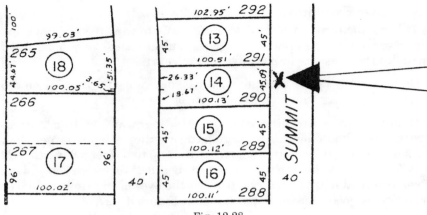

Fig. 12-28

These exhibits are usually drawn on art board rather than flexible material so the attorney does not have a difficult problem setting it up or having it roll up or fold over on him.

Visibility Profiles and Maps

This type of map work is used a great deal for forest fire control to determine the most advantageous positions for lookout posts; the highest peak does not necessarily give the maximum visible area coverage.

The location of shaded areas is another concern to the forestry department so by taking the sun's declination and applying it to profiles of the mountains, *those* areas can be delineated on a map for study. Also because of the difference in vegetation so determined, problems in watersheds and drainage can be given a better analysis.

The visibility profile is furthermore applicable to the line-of-sight requirement for microwave transmission of signals and messages.

The principle of visible and non-visible areas is illustrated in Figure 12-29 where the line-of-sight line is projected across high points of the profile and those areas below and beyond the peaks to its intercept with the next uphill slope are invisible from the lookout tower.

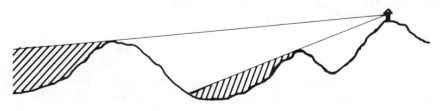

Fig. 12-29

In the same manner that the ground profile was picked off the U.S.G.S. quadrangle sheet and placed on a profile sheet as shown in Figure 7-5, you would likewise draw lines of profiles from a quadrangle sheet, or any other contour map, radiating from the lookout point. Figure 12-30 shows a mountainous area with several peaks to be considered for forestry lookout towers. The highest one at elevation 9638 feet was chosen first so radial lines from it were taken at critical angles to plot profiles for determining those areas that would be observable and those that would not.

A series of profiles ten degrees apart around the compass will give much of the visibility pattern; however, a careful selection of lines through the apparent problem areas, either in addition to, or in lieu of, the ten degree lines will improve the delineation of the pattern.

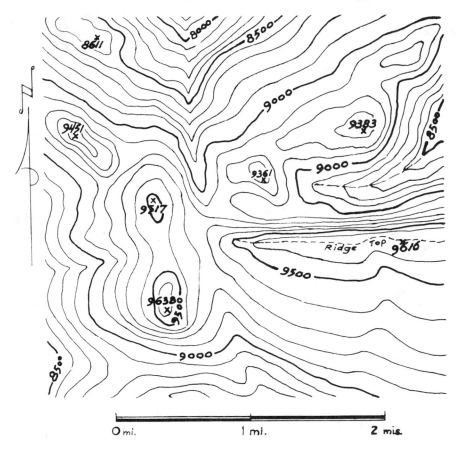

Fig. 12-30

In the situation at hand, the usual line bearing North was drawn through the point under consideration and profiles were taken on lines bearing N 15° 30′ W, N 12° W, N 4° W, N 8° E, N 16° E, N 27° E, N 37° 30′ E, N 42° 30° E and N 48° 30′ E. The ridge above the 9600-foot level

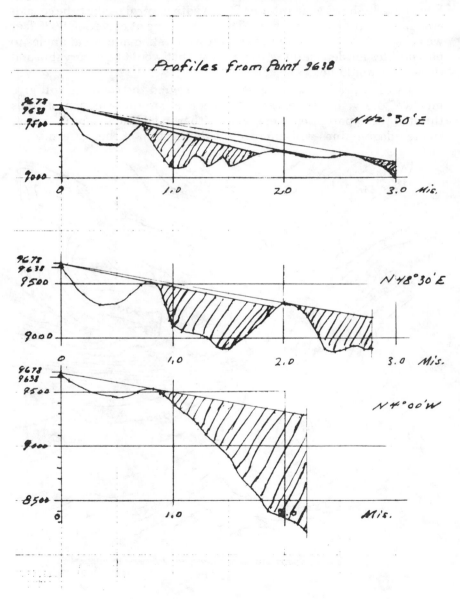

Fig. 12-31

obviously cut off any further view toward the northeast and it is also obvious that the view is good in the sector south of the ridge and to the south, west and west northwest. Profiles on three of the above mentioned bearings are shown in Figure 12-31. Note that the lines of observation were taken from an elevation 40 feet above the ground as representing eye level in the lookout tower.

From the entire set of the above profiles, the pattern of non-visibility was outlined and shaded as shown in Figure 12-32. One conclusion to be reached here is that the tallest peak is not necessarily the best one.

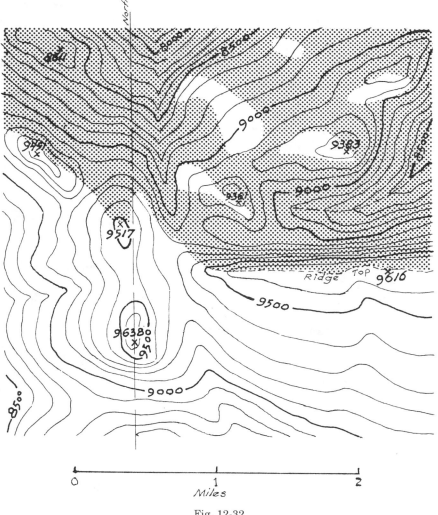

Fig. 12-32

EXERCISE12-4: Draw a series of profiles radiating from Elevation
9616 on the Ridge Top shown in Figure 12-30 and determine the
pattern of non-visibility in this area. From these profiles, mark
the areas which are below the line of sight onto the contour map
in Figure 12-30 and shade them. Is there more, or less, shade
area from this point than the point used in Figure 12-32? Which
one is better?

*   *   *

PROBLEMS:
1. Name 3 types of description by which the transfer of property is
   accomplished through reference to it on a deed.
2. What are the *first* 2 items to be acquired before starting the
   Tentative Tract Map?
3. Name 4 types of information usually shown inside the boundary of
   the proposed subdivision.
4. What names are to appear on the tentative map?
5. Name 10 items of information which may be required to be shown
   on the tentative map pertaining to the tract.
6. Name 4 certificates necessary to the final map.
7. Name 5 additional items usually found on the title page.
8. For a condominium project, name 10 items included in the Condominium Plan Package.
9. Describe a "Deed" Map.
10. Name 4 essential elements to be shown on an Agreement-Line
    Map.
11. Name 3 kinds of Lease Maps.
12. Explain the 2 different types of oil leases for which maps may be
    drawn.
13. What are the 2 types of right-of-way maps?
14. What is an A.L.T.A. Survey Map?
15. Describe the use of visibility profiles.

Chapter 13

Geographic Control

Geographic Control

Whereas we have been discussing the development of surveys, plans, profiles, etc. within limited areas of the plane surface of the earth which are not extensive enough to create, nor require the consideration of, distortion due to the curvature of the earth, the development of maps on a flat surface covering large areas of the rounded surface of the earth presents different problems and is a very special world of cartography.

Forms

In this development, distortions occur because of the differences in viewing the two surfaces. Some of the general forms of projections that have been used are orthographic, stereographic, gnomonic, conic, polyconic, conformal and equal area. Well known developments of large area maps are based on the following particular types of projections: Lambert Conformal Conic, Lambert Azimuthal Equal-area, Albert Conical Equal-area, Bonne Conic, Mercator Equal-area (also known as Sinusoidal), Hassler Polyconic, National Geographic Society Equal-area Polyconic, and Deetz Transverse Polyconic. Perhaps the most used one is the Mercator conformal cylindrical projection, not only for oceanic navigation all over the world, but also for aeronautical charts of the United States Air Force and the British R.A.F. Subsequently, the Transverse Mercator projection (also known as the Gauss conformal) became more desirable for great circle routes over long distance continental and intercontinental flights.

Conformality

One of the desirable attributes to be attained in converting the round surface to a flat surface representation is conformality so that the properties of

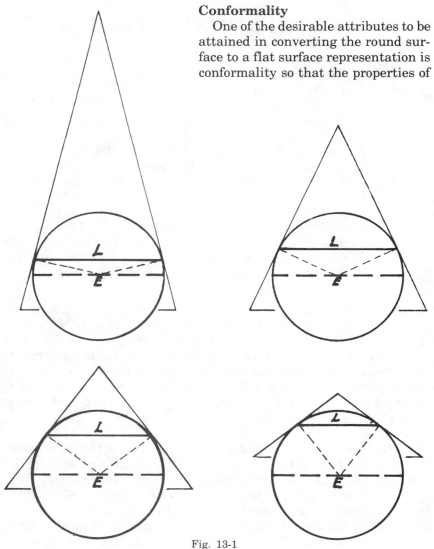

Fig. 13-1

the areas transferred will keep their original forms. To implement this, all angles between intersecting lines and/or curves are preserved.

Polyconic Projection

One type of resemblance has been obtained through the application of the cone to the sphere. Figure 13-1 shows different conic angles applicable to the related degrees of latitude at the point of *tangency*. Combining an ascending series of strips taken from successively flatter cones is known as a Polyconic Projection developed by Ferdinand Hassler, the first Superintendent of the Coast and Geodetic Survey, as shown in Figure 13-2 and its extension shown from a polar view in Figure 13.3. The polyconic projection by construction is not conformal. It is obvious from the plan drawing that any outline of an area transferred by the use

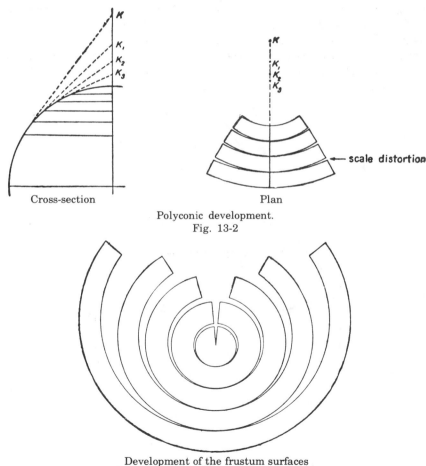

Cross-section   Plan

Polyconic development.
Fig. 13-2

Development of the frustum surfaces
Fig. 13-3

of this form will have distortion increasing in proportion to its distance away from the central meridian; however, Prof. Hassler stated, "This distribution of the projection . . . . . appeared to me the only one applicable to the coast of the United States." It is therefore apparent that this form of transfer should be limited in use to maps of wide latitude but narrow longitudinal extent. To make use of such a method for a wide belt of longitude with a narrow latitudinal extent, the application of this procedure is turned 90° from the poles; it is then called a Transverse Polyconic Projection.

Lambert Zenithal Equal-Area Projection

Another type of projection which preserves the azimuthal relationship and equal areas of those projected from the sphere is known as the Lambert Zenithal (or Azimuthal) Equal-Area Projection. The principle makes use of establishing the central point of the area to be represented as a pole, or zenith point, in the center of the map from which equal great-circle distances in all directions are represented by equal linear distances on the map. It also preserves the property that any portion of the map bears the same ratio to the area represented by it that any other

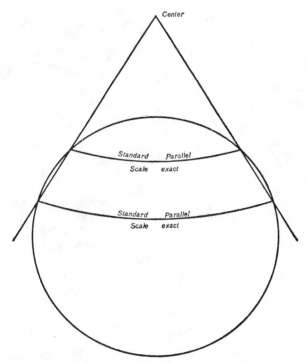

Sphere with intersecting cone for Lambert projection.
Fig. 13-4

portion bears to its corresponding area. This is good for mapping a large area.

Another application of this principle called the Lambert Equal-Area Meridional Projection is used with the center placed on the Equator.

In the first Lambert method above, the exterior limit is a horizon circle while the limit of the second one is a meridional circle.

Disadvantages of the Lambert zenithal projection include the inconvenience of computing coordinates and the plotting of the double system of complex curves of the meridians and parallels along with the intersection of these systems at oblique angles.

Lambert Conformal Conic Projection

Because the application of the cone tangent to the sphere did not cover a wide enough band, another approach was made using the secant principle where the cone was made to cross the circumference of the sphere at two latitudes in contrast with one of the tangent line. See Figure 13-4. This is the basis of the Lambert Conformal Conic Projection whereby the parallels are concentric circles with their center at the point of intersection of the meridians which are coverging straight lines which meet at a common point outside the limits of the map. By using a projection band one-third wider than the width between the latitudes intersected by the cone on a secant basis (one-sixth above and one-sixth below the secant intersects), the scale distortion is no more than 0.01 as shown in Figure 13-5, for standard parallels at 29° and 45°. The scale error here is considerably less than by the use of the Polyconic projection. For state coordinate systems where the smaller dimensions are north and south, the Lambert conformal projection is used and the popular designation for it is the *Lambert grid*.

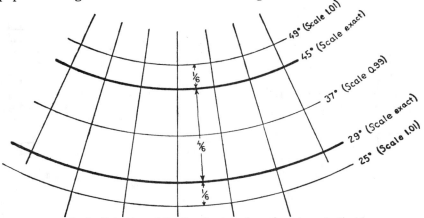

Scale distortion of the Lambert conformal conic projection
with the standard parallels at 29° and 45°.
Fig. 13-5

Mercator Projection

The Mercator Projection uses equally spaced parallel straight lines for the meridians with the parallels of latitude set at right angles to them with their intervals increasing toward each pole in proportion to the lengthening of the parallels with reference to the Equator. A Transverse Mercator Projection is used for state coordinate systems where the zone is narrow east and west and is called a *transverse Mercator grid*.

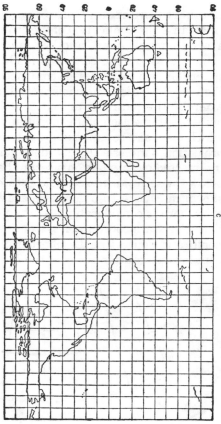

Cylindrical equal-spaced projection.
Fig. 13-7

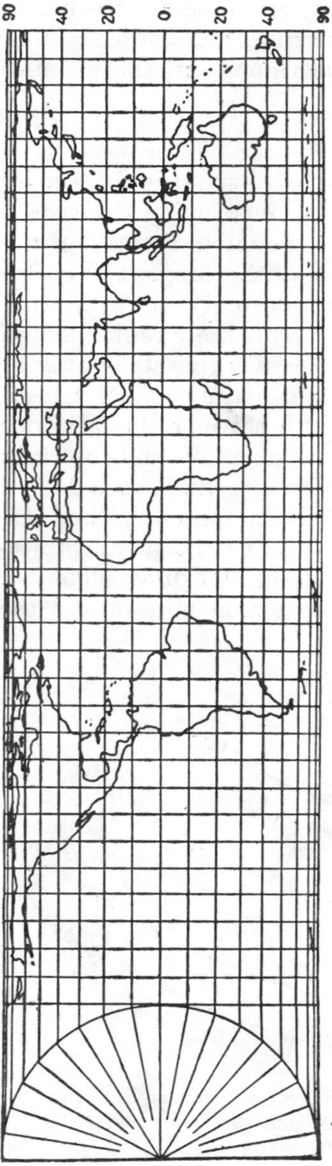

Cylindrical equal-area projection.
Fig. 13-6

Cylindrical Equal-Area Projection

Another type of projection uses a cylinder wrapped around the globe, tangent at the Equator with the meridians parallel north and south. If the parallels are drawn in parallel lines projected from the intersections of radials from the center of a semicircle to its circumference which is tangent to one end of the cylinder, as shown in Figure 13-6, the result is called a cylindrical equal-area projection. This is so because the distance between parallels as they progress from the Equator decrease in proportion to the increase of distances between the meridian in any comparable area.

By spacing the parallels of latitude and the meridians at equal spacing both ways, as shown in Figure 13-7, it reduces the distortion in the polar regions at the expense of the equal area property. This is known as the cylindrical equal-spaced projection.

A compromise of the two foregoing methods is shown in Figure 13-8 where the parallelism is held but the spacing of the meridians and parallels is in the ratio of two to three.

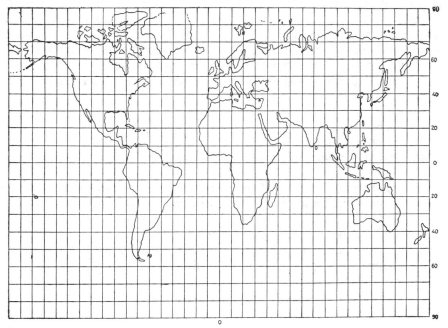

Modified cylindrical equal-spaced projection.
Fig. 13-8

Great Circle Delineation

Another application of the cylinder method is found in the *great circle* delineation. After drawing a line on the globe between two points, rotate it so that the cylinder is tangent to the globe on that line. By tracing a strip of the map onto the cylinder in that area, you will have a route map showing the relation of the great circle line to the meridians and parallels along the way.

Properties

In the effort to obtain reasonable duplication of the facts shown on a round surface by their transfer to a flat surface, there are certain properties that need to be considered in the application of the process:

   a. Correct angular relationship between meridians and parallels.
   b. Trueness of outline or conformality.
   c. Equilvanece of area.
   d. Representation of true distances and azimuths from any given point.
   e. The representation of either a great circle as a straight line or a rhumb line as a straight line.

The purpose for which the finished map will be used will determine the method of its development.

Although this chapter has touched upon a few aspects of map projections as a matter of general information, the application of this subject to nautical and aerial navigation charts involves much more detail and a considerable amount of mathematics. For an in-depth study of this subject, the Author recommends that the reader review both government and private publications concerned with this broad subject of Geographic Control.

PROBLEMS:

   1. What is conformality?
   2. Explain the two principles used in the application of the cone to the sphere.
   3. Is there very much distortion by the use of the conic projection?
   4. Explain the Lambert Zenithal (or Azimuthal) Equal-Area Projection.
   5. Explain the cylindrical projection.
   6. Explain the easy way to map a great circle strip.
   7. Name four properties to consider in the transfer of mapping from a round surface to a flat surface.

Chapter 14

Computerized Drafting

The subject matter of this book would not be complete without some discussion on the progress of computerization in the field of drafting. The outstanding question in the minds of present-day, and future, draftsmen is concerning the displacement of individuals in this field by the computer-plotter. Let us first take an overview of developments in this field and then make an appraisal.

Computer Drafting Operations.

There are two types of computer drafting operations: one is performed on a roller, called a drum plotter, by which you can produce a continuous running sheet of a certain width such as for ownership parcels and/or plan and profile of a long strip of right-of-way. See figure 14-1. Figure 14-2 shows the same type of operation on a vertical stand with microprocessor control and automated selection of up to 8 pens available for choice of colors and line widths. The other is the flatbed for large sized drawings and this is available in either a horizontal or vertical operation. A horizontal type is shown in Figure 14.3. Small table top versions are also available as illustrated in Figure 14.4.

Photo courtesy VTN, Inc.

Fig. 14-1

Photo courtesy of California Computer Products, Inc. (CalComp), Anaheim, CA

Fig. 14.2

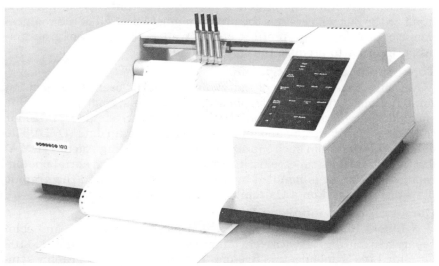

Photo courtesy of California Computer Products, Inc. (CalComp), Anaheim, CA.

Fig. 14.3

Photo courtesy county of Orange, Ca.

Fig. 14.4

Methods of Creating Maps.

There are generally four methods of creating maps through computers: by electrostatic plotting, line plotting, computer output microfilm (COM) and TV screen. The microfilm method is available on 16, 35 and 105 mm film. A hard copy version of the map viewed on the TV screen can be obtained through a printout.

As with all computer programs, however, the output depends upon the input so the one who creates the program must know all of the requirements of drafting maps, profiles, etc., as well as the theory of programming, in order to produce the correct results.

Topographic Analysis System.

Not only have a variety of individual programs been developed but combinations have been interphased to produce special results. One such package is called the Topographic Analysis System (TOPAS) for collecting, manipulating and analyzing digital terrain data (DTD). The data is collected from such sources as Defense Mapping Agency (DMA) digital terrain data tapes, existing topographic maps and aerial photography. The output is handled by a UNIVAC 1100/42 and the graphic plats are produced on a Calcomp 900/1136 drum plotter.

Cartographic Symbolized Data

A Lineal Input System by DMA is used for compilation and color separation and digital data from it is processed through a UNIVAC 1108 Graphic Line Symbolization System. This generates cartographic symbolized data which is produced on the Gerber Plotting System.

Projections

Another Gerber Instrument (model 1232) is a computer-controlled drafting system which offers high precision of 25 μm at the low speed of up to 9.5 cm/s. It can either scribe or ink on a drafting surface or expose a photographic film. The plotter is controlled by a Hewlett Packard 2100A minicomputer. One of its many applications is to generate projections for Albers equal-area and Lambert conformal conic developments.

Several flat-bed drafting machines can print or scribe as much as 20 to 30 inches per second. This is made possible by the use of air-foil supported heads but at this speed, it is necessary to force the drawing ink under pressure. Some of the drum plotters can also reach this speed. From a practical approach, however, it should be pointed out that such rapidity is effective only on relatively long lines and that curves and angle points require slower speed to give proper rendition to the drawing.

Computer Plotting

To present a comparison between the extent of computer drawing and the final product which requires a certain amount of manual drafting, Figure 14-5 shows the subdivision map of Tract No. 11276 as it came off the plotter bed. It was then necessary to manually add information requisite to recording. The final map is shown in Figure 14-6.

By comparing the two, you can see the addition of dimension line arrows, certain lines and notes pertaining to easements etc, adjoining Tract Numbers with their map page references and dash lot lines, exterior street boundary lines which are outside of the calculations for the instant tract, connecting ties to radial bearings etc.

The lettering in the upper left and lower right corners along with the Tract No. and caption are done either with Leroy assist, transparent stick-ons or freehand.

In this case the manual work took as much time as the plotter work. Although the percentages will vary from shop to shop depending on the equipment and personnel, the following tabulation gives a comparison of approximate work (not time) done on computer via manual drafting.

	Computer Plotting	Manual Drafting
Record Map	80%	20%
Exhibit Map	90%	10%
Condominium	100%	
Compiled Record Data	95%	5%
Street Improvements	50%	50%
Grading Plans	50%	50%
Plot Plans including Precise Grading	85%	15%

It is possible to increase the computer percentage by spending more time on the programming but this becomes a matter of judgement concerning the balance between man-hour rates for each classification.

Scanning

Another phase of automation is scanning. One instrument does an excellent job but for all of its precision, it is very slow and highly sensitive to scratches, film or image defects, light pinholes and dust so that the resulting picture may have more than you want on it.

The Sweepnik digitizer produced by the IOM Corp. of Sunnyvale, California is an automatic line follower using a computer controlled laser beam. It reads lines spaced at as much as $250\,\mu$m or as close as $5\,\mu$m increments. The same unit can also plot from digital data by exposing film with a laser beam.

Another system includes a minicomputer, interactive edit display, proofing plotter, magnetic tape unit and digitizing table. The complete operating software part has been developed to support the scaling, paneling, cartographic digitizing and editing, along with the generating of output plat formats.

Marine Data Systems

To assist and enhance nautical charting by implementing a computer-assisted system, the Marine Data Systems Project (MDSP) was established by National Ocean Survey. To obtain an in-house conversion from graphic chart data to digital format in order to establish a nautical chart data bank, a five-table integrated digitized system was put into operation. From this bank data, a laser-raster production plotter creates reproducible negatives of the charts.

Fig. 14.5

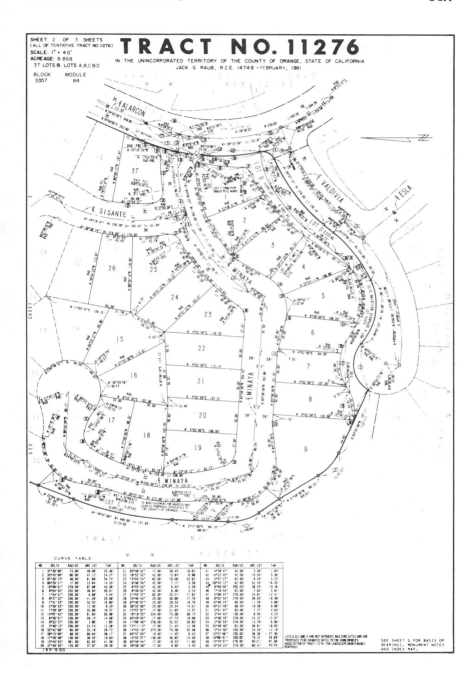

Fig. 14.6

-10 10 A.M.   OCT 07 '80

$11.00
C7

LEE A. BRANCH, County Recorder

CONDOMINIUM PLAN FOR LOT 1 OF

TRACT NO. 10228

IN THE UNINCORPORATED TERRITORY OF THE COUNTY OF ORANGE, STATE OF CALIFORNIA.

JUNE, 1979   JACK G. RAUB   RCE 14749

ALTA FINISTERRA

OWNERSHIP CERTIFICATE AND ACKNOWLEDGEMENT

WE, THE UNDERSIGNED, BEING ALL PARTIES HAVING ANY RECORD TITLE INTEREST IN THE LAND INCLUDED WITHIN THE PROJECT AS SHOWN ON THIS PLAN, DO HEREBY CONSENT TO THE PREPARATION AND RECORDATION OF SAID PLAN, PURSUANT TO THE PROVISIONS OF SECTION 1351 OF THE CIVIL CODE.

MISSION VIEJO COMPANY,
A CORPORATION

MARVIN E. LAWRENCE   VAN STEVENS
SENIOR VICE PRESIDENT   VICE PRESIDENT

STATE OF CALIFORNIA] ss
COUNTY OF ORANGE]

ON THIS 22nd DAY OF August , 1979,
BEFORE ME, Peggy J. Miklaus ,
A NOTARY PUBLIC IN AND FOR SAID COUNTY AND STATE, PERSONALLY APPEARED MARVIN E. LAWRENCE, KNOWN TO ME TO BE SENIOR VICE PRESIDENT AND VAN STEVENS, KNOWN TO ME TO BE VICE PRESIDENT OF THE MISSION VIEJO COMPANY, A CORPORATION, THE CORPORATION THAT EXECUTED THE WITHIN INSTRUMENT AND KNOWN TO ME TO BE THE PERSONS WHO EXECUTED THE WITHIN INSTRUMENT ON BEHALF OF SAID CORPORATION AND THEY ACKNOWLEDGED TO ME THAT SUCH CORPORATION EXECUTED THE SAME.

WITNESS MY HAND AND OFFICIAL SEAL   Peggy J. Miklaus
NOTARY PUBLIC IN AND FOR
SAME COUNTY AND STATE

OFFICIAL SEAL
PEGGY J. MIKLAUS
NOTARY PUBLIC CALIFORNIA
PRINCIPAL OFFICE IN
ORANGE COUNTY
My Commission Expires April 10, 1981

ENGINEER'S CERTIFICATE

I HEREBY CERTIFY THAT I AM A REGISTERED CIVIL ENGINEER OF THE STATE OF CALIFORNIA AND THAT THIS PLAN, CONSISTING OF 9 SHEETS, CORRECTLY REPRESENTS, (1) A TRUE AND COMPLETE SURVEY OF THE PERIMETER OF THE PROJECT I LOT 1, TRACT 10228 I MADE UNDER MY SUPERVISION IN FEBRUARY, 1978, AND (2) THE LOCATIONS OF AIRSPACE AND BUILDINGS TO BE INDICATED.

JACK G. RAUB   RCE 14749

BENCHMARK   3-E-88-71

O.C.S. ALUMINUM CAP IN THE SW'LY PART OF THE INTERSECTION OF TRABUCO ROAD AND MARGUERITE PARKWAY; 88' S'LY FROM THE CENTERLINE OF THE MEDIAN OF "TRABUCO ROAD, 39' W'LY FROM THE CENTERLINE OF THE MEDIAN OF MARGUERITE PARKWAY, SET IN THE TOP NE'LY CORNER OF A 5.5'X 6.5' UNDERGROUND CONCRETE VAULT, LEVEL WITH THE SIDEWALK
LEV. 1971, ADJ. 1970   594.157 ELEVATION

INDEX MAP

GENERAL NOTES

1. THE COMMON AREAS OF THIS PHASE 3 OF THE CONDOMINIUM PROJECT CONSISTS OF THE LAND AND REAL PROPERTY INCLUDED WITHIN LOT 1 OF TRACT 10228, RECORDED IN BOOK _____ PAGES _____ THROUGH _____ OF MISCELLANEOUS MAPS, IN THE OFFICE OF THE COUNTY RECORDER OF THE COUNTY OF ORANGE, CALIFORNIA, AS SHOWN ON SHEETS 2, 3 AND 4, HEREIN, EXCEPT THEREFROM UNITS 1 THROUGH 58, AS SHOWN AND DESCRIBED HEREIN, AND AS FURTHER DEFINED IN THE DECLARATION RECORDED CONCURRENTLY HEREWITH.

2. THE LIVING ELEMENTS, GARAGE AND CARPORT ELEMENTS OF THE UNITS ARE NUMBERED 1 THROUGH 58.

3. FLOOR PLAN DESIGNATIONS ARE: PLAN 1, PLAN 1L, PLAN 1AL, PLAN 2, PLAN 3 AND PLAN 3L. "R" FOLLOWING ANY PLAN NUMBER DESIGNATES REVERSE FLOOR PLAN.

4. ABBREVIATIONS: L.E. - LIVING ELEMENT, B - BALCONY, P - PATIO, G - GARAGE, C - CARPORT, F.F.EL. - FINISH FLOOR ELEVATION, TYP - TYPICAL, A - ATRIUM.

5. UNITS COMPRISING PLANS 1, 1R, 1L, 1LR, 1AL, 1ALR, 3, 3R, 3L AND 3LR ARE COMPOSED OF FOUR ELEMENTS: LIVING ELEMENT, BALCONY OR PATIO, GARAGE AND CARPORT. UNITS COMPRISING PLANS 2 AND 2R ARE COMPOSED OF FOUR ELEMENTS: LIVING ELEMENT, BALCONY, ATRIUM AND GARAGE.

6. THE BOUNDARIES OF THE LIVING ELEMENT ARE THE INTERIOR FINISHED SURFACES OF THE PERIMETER WALLS, FLOORS, CEILINGS, FIREPLACE, WINDOWS AND DOORS, AS SHOWN HEREIN, AND AS FURTHER DEFINED IN THE DECLARATION RECORDED CONCURRENTLY HEREWITH. THE HORIZONTAL AND VERTICAL DIMENSIONS OF THE LIVING ELEMENT ARE AS SHOWN ON SHEETS 5 THROUGH 7, INCLUSIVE.

7. THE BOUNDARIES OF THE BALCONY, PATIO AND ATRIUM ELEMENTS ARE THE SURFACES OF THE FLOORS TO THE EXTERIOR BOUNDARY LINES, AS SHOWN HEREIN, AND AS FURTHER DEFINED IN THE DECLARATION RECORDED CONCURRENTLY HEREWITH. THE HORIZONTAL AND VERTICAL DIMENSIONS OF THE BALCONY AND PATIO ELEMENTS ARE AS SHOWN ON SHEETS 5 THROUGH 8, INCLUSIVE. THE HORIZONTAL AND VERTICAL DIMENSIONS OF THE ATRIUM ELEMENT ARE AS SHOWN ON SHEET 7.

8. THE BOUNDARIES OF THE GARAGE ELEMENT ARE THE INTERIOR SURFACES OF THE FLOORS AND CEILINGS TO THE PERIMETER WALLS THEREOF, AS SHOWN HEREIN, AND AS FURTHER DEFINED IN THE DECLARATION RECORDED CONCURRENTLY HEREWITH. THE HORIZONTAL AND VERTICAL DIMENSIONS OF THE GARAGE ELEMENT ARE AS SHOWN ON SHEETS 7 THROUGH 9, INCLUSIVE.

9. THE BOUNDARIES OF THE CARPORT ELEMENT ARE THE INTERIOR SURFACES OF THE FLOORS AND CEILINGS TO THE EXTERIOR BOUNDARY LINES, AS SHOWN HEREIN, AND AS FURTHER DEFINED IN THE DECLARATION RECORDED CONCURRENTLY HEREWITH. THE HORIZONTAL AND VERTICAL DIMENSIONS OF THE CARPORT ELEMENT ARE AS SHOWN ON SHEET 9.

10. THE FOLLOWING ARE NOT PART OF A UNIT: BEARING WALLS; COLUMNS; SOFFITS; VERTICAL SUPPORTS; HORIZONTAL SUPPORTS; FLOORS; CEILINGS; ROOFS; FOUNDATIONS; BEAMS; PATIO AND BALCONY WALLS; FENCING AND LANDSCAPING; CENTRAL SERVICES; PIPES; DUCTS; FLUES; CHIMNEYS; WIRES; AND OTHER UTILITY INSTALLATIONS, WHEREVER LOCATED, EXCEPT THE OUTLETS THEREOF WHEN LOCATED WITHIN A UNIT, AND AS FURTHER DEFINED IN THE DECLARATION RECORDED CONCURRENTLY HEREWITH.

11. THE PLANE ANGLE BETWEEN THE BOUNDARY LINES OF EACH UNIT IS 90 DEGREES.

12. BUILDING CONTROL DIMENSIONS SHOWN ON SHEETS 3 AND 4 ARE MEASURED FROM THE CENTERLINE OF COMMON WALLS, THE FOUNDATION CORNERS, THE FOUNDATION SIDELINES, OR THEIR PROJECTIONS.

BASIS OF BEARINGS

THE BEARINGS SHOWN HEREON ARE BASED ON THE CENTERLINE OF MARGUERITE PARKWAY BEING N 9° 41' 00" W, AS SHOWN ON THE MAP OF TRACT NO. 9073 RECORDED IN BOOK 373, PAGES 1 THROUGH 10, OF MISCELLANEOUS MAPS, IN THE OFFICE OF THE COUNTY RECORDER OF THE COUNTY OF ORANGE, CALIFORNIA.

TYPICAL BUILDING DETAIL

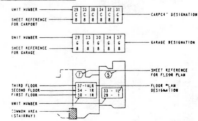

Jack G. Raub Company
Engineering & Planning

JN. 01-22-014

Fig. 14.7

Fig. 14.8

Digital Planimetric Compiler

The U.S. Army Engineer Topographic Laboratories (USAETL) has developed a Digital Planimetric Compiler (DPC) for digitizing planimetric data from orthophotography. From film transparencies, a tracing cursor leaves a mark of the features as they are traced. Incremental vector codes are generated by every 5 milli-in. of cursor movement. The system is made up of five digitizers on-line to a single minicomputer.

A similar instrument is the digitizer which, instead of tracing the line, pinpoints board coordinates into a command post for planned assignment and retrieval. A more significant name for "board coordinates," however, would be "drawing board reference numbers" because *they* are not coordinates as applied to survey work.

From a map or a drawing of survey lines and corners, the dot in the center of the tracing glass can be placed on critical points and the digitizer is commanded to record the position in terms of the board numbers. By command to the plotter, lines are subsequently drawn between these points including curves.

If a known control, such as Lambert State Plane Coordinate System, has been established on any of the map or survey points, it can be also inputted with the command for it to be the controlling factor in lieu of the board numbers. A typical setup of the board and recorder is shown in Figure 14-9.

Computer Lettering

There have been any number of programs written to generate lettering and symbols for automated drafting, the size of which can be easily expanded or contracted by command orders to the output. The computer-controller can also be given area limits within which to place certain data and if it is more than the allotted space will accommodate, the computer will store it in memory and assign a reference number for subsequent tabular printout.

A new Cathode Ray Tube (CRT) plotting head for lettering and symbols is being developed which is five inches in diameter and attached to the plotter carriage. After reaching the desired position, the CRT, controlled by its own minicomputer, generates the required symbol and exposes it onto photographic film at the plotting surface. Because the characters are written and stored in raster format, a raster scanner can be used to update any stored data. Digitized fonts of characters for various systems will be available.

It is not a question of, "*can* the computer plotter do any kind of lettering desired?" it is again a matter of the time and programming involved versus other methods.

Photo courtesy county of Orange, Ca.

Fig. 14-9

Where the same matter is required time after time with perhaps a minor change here and there, the use of a programmed printout can be justified. An example is shown in Figure 14-7.

Although photo-typo-graphics is not computerization, it should be considered here for comparison. This kind of typography is faster, cleaner and easier to handle than the former hard-metal system. Its versatility and speed in adapting letters, figures and some symbols to size and spacing is a big plus in its favor. Each font is reproduced on a film in a standard size card which is placed in the console in line with a variable-focus lens which by adjustment renders the size of the letter desired. Another operation determines the spacing between words. The printout can be on either transparent film or opaque paper for sticking onto the overall sheet.

The title sheet for Tract No 11276 shown in Figure 14-8 was done by this method except for the symbols, notary seal, Index Map and heavy block lettering.

The choice of method is a matter of economics and depends upon the combination of equipment and technical personnel available in each office.

As impressive as this information appears, does it really mean that draftsmen will be supplanted by computer-controlled plotters? Even when one considers the very high cost of these instruments, it follows that those offices which include drafting on engineering and surveying projects in any appreciable volume will make valuable use of these new sophisticated robots. Furthermore (referring back to the comparative table on page 14.5) draftspersons are still required to complete the drawing.

Appraisal

Most important of all, however, is the need for individuals who *understand* property boundary requirements, the value of monuments relative to controlling factors, field note interpretation, mapping relative to legal descriptions, coordinates as applied to field and office work versus their legal determination, x, y and z axis relationships, earthwork calculations, improvement plans, subdivision maps, etc., etc. It is not just a matter of doing any one or all of these things, it is a matter of knowing *how* to relate them and what to do in the case of any problems.

The machine cannot make judgment decisions and this is where the individual who understands the ramifications of survey drafting, and knows what to do, possesses an inherent value far above that of the machine; consequently, there is a place for those of you who have studied and learned such detailed knowledge in the field of drafting.

Problems.

1. Name the two types of computer controlled drafting machine.
2. Name four methods of creating maps through computers.
3. Name two well known projections that can be generated by a computer-controlled plotter.
4. Describe two systems for computer-produced letters and symbols printing.
5. Describe the principle of the digitizer.
6. Do you believe that the computerized drafting machine will replace the human draftsman?

Appendix

Abbreviations . . .

There are hundreds of abbreviations used in surveying and engineering and almost every office has some of its own particular designations for special applications. The following is a cross section of many of the more commonly used shorthand expressions.

Above	Abv.	Average	Av. or Aver.
Access	Acc.		
Acoustic	Acst.	Azimuth	Az.
Acres	Ac.	Backsight	B.S.
Addendum	ADD	Base line or building line	B.L.
Adjoining or adjusted	Adj.		
Air condition	Air Cond.	Bearing	Brg.
Altitude	Alt.	Beginning of curve	BC
Anchor bolt	AB	Beginning of vertical curve	BVC
Angle	Ang.		
Antenna	Ant.	Bench mark	BM
Architecture	Arch.	Bevel	Bev.
Area	A.	Block	Blk.
Asphalt concrete pavement	A.C. Pav.	Boulevard	Blvd.
		Boundary	Bdry.
Asymmetric	Asym.	Brick	Brk.
Automatic	Auto.	Bridge	Brdg.
Auxiliary	Aux.	Building	Bldg.
Avenue	Ave.	By (between dimensions)	X

Calculate	Calc.	Diagonal	Diag.
Capacity	Cap.	Diameter	Dia.
Cast iron pipe or corrugated iron pipe	C.I.P.	Difference	Diff.
Catch basin	C.B.	Disintegrated granite	D.G.
Ceiling	Clg.	Distance or District	Dist.
Cement	Cem.	District map	D.M.
Center	Ctr.	Does not apply	DA
Center line	CL or ₵	Door	Dr.
Center to center	C to C	Douglas fir	D.F.
Chain	Ch.	Down	Dn.
Channel	Chan.	Drain or Drive	Dr.
Chiseled cross	Ch.X.	Drawing	Dwg.
Chord	Chd.	Drill hole	D.H.
Circle	Cir.	Drop manhole	D.M.H.
City Engineer	C.E.	Each	Ea.
City Engineer's Field Book	C.E.F.B.	Easement	Esmt.
Clear	Clr.	East	E.
Column	Col.	Electric	Elect.
Common	Com.	Elevation	EL or Elev.
Complete	Comp.	End of curve	E.C.
Concave	Concv.	End of return	E.R.
Concrete	Conc.	End of vertical curve	E.V.C.
Concrete block	CCB	Equipment	Eqpmt.
Concrete floor	CCF	Except	EXC
Continued	Cont.	Existing	EX or Exst.
Contour	Cntr.	Facing	Fcg.
Copper tack	C.T.	Field book	F.B.
Corner	Cor.	Fill	F.
Corrugated metal pipe	C.M.P.	Fire alarm box	FABX
County	Co.	Fire door	FDR or F.Dr.
County Surveyor	C.S.	Fire hydrant	FH
County Surveyor's Field Book	C.S.F.B.	Flood control	F.C.
Cross section	X Sect.	Floor	Fl.
Cubic centimeter	cc	Floor drain	FD
Cubic foot	cu. ft.	Flow line	F.L.
Culvert	Culv.	Flush	Fl.
Curb face	C.F.	Footing	Ftg.
Curb return	C.R.	Foresight or finished surface	F.S.
Deflection	Def.	Foundation	Fdn.
Degree	Deg.	Fractional	Frac.
Degree of curvature	D.		

Front	Fr.	Lineal foot or feet	Lin. ft.
Galvanized	Galv.	Locate or location	Loc.
Glass	Gl.	Longitude	Long.
Government	Govt.	Lower low water line	LLWL
Grade	Gr.	Low water line	LWL
Grade change	G.C.	Lumber	Lbr.
Grade line or ground line	G.L.	Magnetic	Mag.
Gravel	Gvl.	Main	Mn.
Ground	Gnd.	Major	Maj.
Head chainman or house connection	H.C.	Manhole	M.H.
Height of instrument	H.I.	Map	Mp.
High water line	HWL	Map book	M.B.
High water mark	HWM	Mark	Mk.
Highway	HWY	Masonry	Msnry.
Horizontal	Hor.	Material	Matl.
Horizontal center line	HCL	Maximum	Max.
Hose bib	HB	Mean high tide	MHT
Hot water	HW	Mean higher high water	MHHW
Improvement	Impov. or impvmt.	Mean high water	MHW
		Mean low tide	MLT
Incomplete	Incomp.	Mean lower low water	MLLW
Incorrect	Incor.	Mean low water	MLW
Independent	Indt.	Mean sea level	MSL
Industrial	Ind.	Measure	Meas.
Inlet	Inl.	Meter	Mtr.
Inside diameter	ID	Middle of curve	M.C.
Installation	Instal.	Middle or midway	Mid.
Instrument	Inst.	Midordinate	M.O.
Intake	Intk.	Midpoint	Mdpt.
Interior	Intr.	Minimum or minute	Min.
Intersection	Inters.	Miscellaneous	Misc.
Iron pipe	IP	Miscellaneous records	M.R.
Junction chamber	J.C.	Monument	Mon.
Junction structure	J.S.	Nail and tin	N. & T.
Lamp hole	L.H.	National Geologic Survey	N.G.S.
Landmark	Ld. Mk.	National Oceanic Survey	N.O.S.
Land Surveyor	L.S.	Natural	Nat.
Latitude	Lat.	Negative	Neg.
Layout	Lyt.	Nonreinforced concrete pipe	NRCP
Lead and tack	L. & T.	Normal	Norm.
Length of arc	L.	North	N
Level	Lev.	Not a part of	NAP

Not applicable	NA
Not to scale	NTS
Official records	O.R.
On center	OC
Ordinance	Ord.
Ordinary high water mark	OHWM
Original	Orig.
Outlet	Out.
Outside diameter	OD
Overhead	Ovhd.
Parallel	Par. or //
Parkway	Pkwy.
Part	Pt.
Party Chief	P.C.
Paved or paving	Pav.
Pavement	Pvmt.
Perpendicular	Perp.
Pipeline	PPLN or Pi.Li.
Place	Pla.
Plan view	PV
Plaster	Plas.
Plus or minus	PORM or ±
Point	Pt.
Point of beginning	POB
Point of compound curve	PCC
Point of curve	PC
Point of intersection	PI
Point of reverse curve	PRC
Point of tangency	PT
Point on curve	POC
Point on line	POL
Point on semi tangent	POST
Point on tangent	POT
Positive	Pos.
Power	Pwr.
Power and lighting distribution	P&L Distr.
Power plant	Pwr. Pl.
Power pole	P.P.
Power supply	Pwr. Sply.

Precise bench mark	P.B.M.
Primary	Pri.
Private	Priv.
Property in question	P.I.Q.
Property line	P.L. or ₽
Public address	PA
Pump	Pmp.
Radius or radial	Rad.
Railroad	RR
Railway	Ry.
Rancho	Ro.
Rear chainman	R.C.
Record	Rec.
Redhead	R.H.
Red wood	R.W.
Reference	Ref.
Reference line	R.L.
Reference mark	R.M.
Reference point	R.P.
Reinforced concrete	RC
Reinforced concrete pipe	RCP
Relocated	Reloc.
Removable or removed	Rem.
Reservoir	RSVR or Res.
Retaining wall	Ret. wall
Reverse or revised	RVS or Rev.
Right angle	RTANG or Rt. ∠
Right of way	R/W
Road	Rd.
Rock and oil	R. & O.
Room	Rm.
Round	Rnd.
Sanitary sewer	S.S.
Sea level	SL
Section	Sect.
Semi-tangent	S.T.
Shoulder	Shldr.
Signal	Sig.
Single	SGL or Sin.
Slope	Slp.

Sounding	Sdg.	Traffic signal	T.S.
South or sewer	S	Transition	Trans.
Specifications	Spec.	Traverse line	T.L.
Spike	Spk.	Turning point	T.P.
Spike and tin	S. & T.	Typical	Typ.
Spike and washer	S. & W.	Underground	UGND
Stairway	Stwy.	U.S. Coast &	U.S.C. &
Stake	Stk.	Geodetic Survey	G.S.
Standard	Std.	U.S. Corps of	
Standard bench mark disk	S.B.M.D.	Engineers	U.S.C.E.
		U.S. Geological Survey	U.S.G.S.
Standard survey monument	S.S.M.	Unmarked	Unmkd.
Standpipe	S.P.	Utility	Util.
Station	Sta.	Vacant	Vac.
Steel	Stl.	Valley	Val.
Storage	Stor.	Valve	V
Storeroom	St. Rm.	Valve box	VB
Storm drain	S.D.	Vent pipe	VP
Straight grade	Str. Gr.	Vertical	Vert.
Structure	Struct.	Vertical center line	VCL or Ver. ₵
Substation	Sub. Sta.	Vertical curve	VC
Substructure	Sub. Str.	Vitrified clay pipe	VCP
Survey	Sur.	Void	VD
System	Sys.	Volume	Vol.
Tack or tank	Tk.	Walk	Wk.
Tangent	Tan.	Warehouse	Whse.
Telegraph	TLG	Waste pipe	WP
Telephone	Tel.	Water heater	WH
Telephone pole	T.P.	Water line	WL
Temperature or temporary	Temp.	Water meter	WM
Temporary bench mark	T.B.M.	Water pump	WP
		Weight	Wt.
Terminus	Term.	West	W.
Terrace	Ter.	Window	Win.
Test hole	T.H.	Wiring	Wrg.
Thick	Thk.	Witness corner	W.C.
Total	Tot.	Witness post	W.P.
Tract	Tr.	Wood	Wd.

Topographic Symbols

Courtesy of U.S. Army
and U.S.C. & G.S.

Primary highway, hard surface . . . .

Secondary highway, hard surface . . . .

Light-duty road, hard or improved surface . . . .

Unimproved road . . . .

Road under construction, alinement known . . . .

Proposed road . . . .

Dual highway, dividing strip 25 feet or less . . . .

Dual highway, dividing strip exceeding 25 feet . . . .

Trail . . . .

Dual Highway.

Consists of two or more lanes on each side of a physical separation such as a parkway. The number of lanes is indicated by labeling.

Hard-Surface, Heavy-Duty Road.

Labeling indicates the number of lanes.

Hard-Surface, Heavy-Duty Road.
Unless otherwise labeled, the road is two lanes wide.

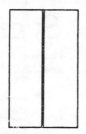 

Hard-Surface, Medium-Duty Road.
Unless otherwise labeled, the road is two lanes wide.

Hard-Surface, All-Weather Road, Two or More Lanes Wide.

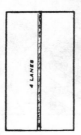 

Hard-Surface, All-Weather Road, One Lane Wide.

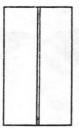 

Light-Surface, All-Weather Road, Two or More Lanes Wide.

Light-Surface, All-Weather Road, One Lane Wide.

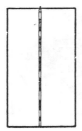 

Improved, Light-Duty Road.

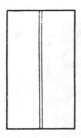 

Unimproved Dirt-Road.

Road Under Construction.   Only roads on which construction is actually under way are shown.

Point of Change in Number of Lanes of Extra-Width Road.

Streets in Developed Areas.   Streets are shown to agree with the cultural density and the scale of the map. Normally, streets are symbolized alike regardless of construction. If width permits, a street is shown to scale. Alleys are not shown.

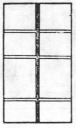   

Route Marker.   (a) Federal or national. (b) State, province or equivalent.

Main Road.

Secondary Road.

Other Road.

Track.

Trail.

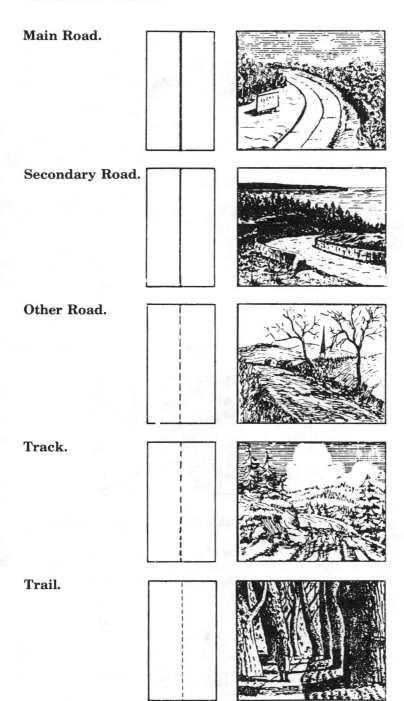

Railroad: single track and multiple track . . . .

Railroads in juxtaposition . . . .

Narrow gage: single track and multiple track . . . .

Railroad in street and carline . . . .

Bridge: road and railroad . . . .

Drawbridge: road and railroad . . . .

Footbridge . . . .

Tunnel: road and railroad . . . .

Overpass and underpass . . . .

Small masonry or concrete dam . . . .

Dam with lock . . . .

Dam with road . . . .

Canal with lock . . . .

Single-Track Railroad, in Operation.
(a) Normal or broad gage. (b) Narrow gage. Broad and narrow gage railroads are labeled as to gage.

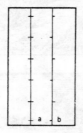

Double- or Multiple-Track Railroad, in Operation. Railroad is double-track if not otherwise labeled.
(a) Normal or broad gage. (B) Narrow gage. (c) Standard gage (for use in United States only). Broad gage railroads are labeled as to gage.

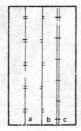

Railroads in Juxtaposition.   Railroads of different ownership closely parallel.

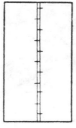 

Single-Track Railroad, Nonoperating.   Labeling indicates whether railroad is abandoned, destroyed, or under construction.
(a) Normal or broad gage. (b) Narrow gage. Broad and narrow gage railroads are labeled as to gage.

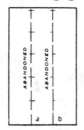 

Railroad Siding.   Included are tracks for passing, storage, and loading and unloading of passengers or freight.

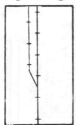 

Railroad Yard.   The limiting sidings indicate the correct shape of the yard. Lines inside the outline are symbolic and do not show the correct number of sidings.

Turntable. A turntable is not drawn to scale. It is usually omitted in congested areas.

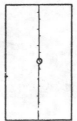 

Railroad Station. Within built-up areas, stations are shown only when they are significant as landmarks. If the building is identifiable, it appears in correct location. (a) Position known. (b) Position unknown.

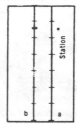 

Railroad Snowshed.

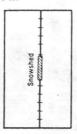 

Carline. Carlines are shown only in open areas; they are omitted in built-up areas, (a) Single. (b) Double.

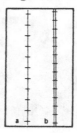 

Horizontal and vertical control station:

    Tablet, spirit level elevation . . . .   BM △ 5653

    Other recoverable mark, spirit level elevation . . . .   △ 5455

Horizontal control station: tablet, vertical angle elevation VABM △ *95/9*

    Any recoverable mark, vertical angle or checked elevation   △*3775*

Vertical control station: tablet, spirit level elevation . . . . .   BM ✕ 957

    Other recoverable mark, spirit level elevation . . . .   ✕ 954

Spot elevation . . . . ✕ *7369*   **✕ *7369***

Water elevation *670*   ***670***

Horizontal Control Point.   The symbol represents a described horizontal control point which is marked on the ground and which was established by triangulation or traverse of third or higher order accuracy.

Monumented Bench Mark.   The symbol represents a described vertical control point which is marked by a tablet on the ground and which was established by survey methods of third or higher order accuracy. (a) and (b) are alternate symbols.

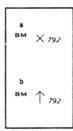

Monumented Bench Mark At Horizontal Control Point. The symbol represents a described control point which is marked on the ground and whose horizontal and vertical positions were established by survey methods of third or higher order accuracy.

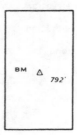

Non-monumented Bench Mark. (sometimes called temporary, supplementary, or intermediate). The symbol represents a described vertical control point which was established by survey methods of third or higher order accuracy. The point is usually recoverable. The mark does not bear a tablet.

Checked Spot Elevation. The symbol represents an elevation established by closed lines, including spirit level, stadia, and vertical angle methods.
(a) Identifiable point. (b) Unidentifiable point.

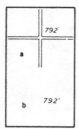

Unchecked Spot Elevation. The symbol represents an elevation determined by unchecked field surveys such as side shots on stadia lines, unchecked vertical angle or precision altimetry, or by repeated photogrammetric readings. An un-checked spot elevation is not as reliable as checked spot elevations.
(a) Identifiable point. (b) Unidentifiable point.

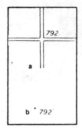

Buildings (dwelling, place of employment, etc.)	
School, church, and cemetery	⌖ ⊹ ┼ Cem
Buildings (barn, warehouse, etc.)	▫▭
Power transmission line with located metal tower	·----·----▫-----·
Telephone line, pipeline, etc. (labeled as to type)	———————·
Wells other than water (labeled as to type)	○Oil ○Gas
Tanks: oil, water, etc. (labeled only if water)	▫●● ⊘Water
Located or landmark object; windmill	○ ⚥
Open pit, mine, or quarry; prospect	⌄ x
Shaft and tunnel entrance	▪ Y

Power-Transmission Line. These are shown only when they are landmark features in areas of sparse cultural development. They are seldom shown along roads and railroads. Voltage is not indicated.

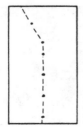

Telephone and Telegraph Lines. These are shown only when they are landmark features in areas of sparse cultural development. They are usually omitted along roads or railroads.

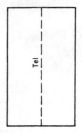

Pipeline. This includes only those pipelines not used for water which are landmark features in areas of sparse culture. They are omitted in developed areas. No effort is made to show pipelines as a continuous feature and only landmark parts are shown.

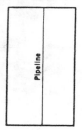

Underground Pipeline. The symbol represents underground pipelines which are obvious from cleared rights-of-way, ground scars, or levee-like mounds.

Aqueduct. The symbol represents a conduit used for carrying water. It may be either an open or closed canal. Water pipelines are symbolized by the aqueduct symbol.

Landmark or Located Object. A feature is a landmark when it is visible from a distance. Landmarks include towers, chimneys, air beacons, monuments, and similar features. Labeling indicates the nature of the object.

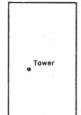 

Perennial streams		Intermittent streams	
Elevated aqueduct		Aqueduct tunnel	
Water well and spring		Glacier	
Small rapids		Small falls	
Large rapids		Large falls	
Intermittent lake		Dry lake bed	
Foreshore flat		Rock or coral reef	
Sounding, depth curve	10	Piling or dolphin	o
Exposed wreck		Sunken wreck	
Rock, bare or awash; dangerous to navigation			
Marsh (swamp)		Submerged marsh	
Wooded marsh		Mangrove	
Woods or brushwood		Orchard	
Vineyard		Scrub	

Narrow Perennial Stream.
(a) Surveyed. (b) Unsurveyed.

Wide Perennial Stream.
(a) Surveyed. (b) Unsurveyed.

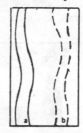 

Intermittent Stream.

Glacial Moraine.

Shoreline. The mean high or normal water line is the shoreline.
(a) Definite. (b) Indefinite or unsurveyed.

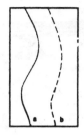 

Perennial Lake or Pond.
 (a) Definite shoreline. (b) Indefinite or unsurveyed shoreline.

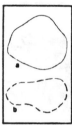 

Intermittent Lake or Pond.

Dry or Cyclical Lake or Pond.

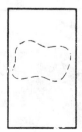 

Marsh or Swamp. No distinction is made between fresh and salt marshes.

Coastal Marsh in Tidal Waters. The shoreline is drawn as the water side limits of the marsh.

Coastal Marsh in Nontidal Waters. The shoreline is drawn as the true shoreline.

Vineyard. Vine growths which are not perennial are omitted. No indication as to the type of growth is given.

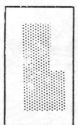 

Woods or Brushwood.   Any perennial vegetation of enough stand or height or which is thick enough to be a serious obstacle to free passage is classified as woods or brushwood. No distinction is made between woods and brushwood or between different types of vegetation. Rendition is made with a wash of green color.

Scrub.   Scrub growth includes cactus, stunted shrubs, sagebrush, mesquite, and similar plants of low growth which present an obstacle to free passage or which serve as landmarks in desert areas. Rendition is stippled in green color.

Orchard or Plantation.   An area of orchards or plantations usually consists of rows of evenly spaced trees. The type of growth is indicated except when it is of the common fruit variety, such as apple, orange, pear, or the like.

Fill	↘↙↙	Cut	
Levee		Levee with road	
Mine dump		Wash	
Tailings		Tailings pond	
Shifting sand or dunes		Intricate surface	
Sand area		Gravel beach	

Strip Mine, Tailings Pile, Mine Dump.

Open-Pit Mine or Quarry.   This feature is usually omitted in areas of dense culture. In foreign areas all types of mines are shown by this symbol. Whenever possible, the appropriate labeling supplements the symbol indicating the material mined.

Coal

Mine Tunnel or Shaft.

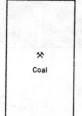

Crescent Dunes.

Lateral Dunes.

Wet Sand.

Sand Beach.

Boundaries: National . . . .

State . . . .

County, parish, municipio . . . .

Civil township, precinct, town, barrio . . . .

Incorporated city, village, town, hamlet . . . .

Reservation, National or State . . . .

Small park, cemetery, airport, etc . . . .

Land grant . . . .

Township or range line, United States land survey . . . . . .

Township or range line, approximate location . . . .

Section line, United States land survey . . . .

Section line, approximate location . . . .

Township line, not United States land survey . . . .

Section line, not United States land survey . . . .

Found corner: section and closing . . . .

Boundary monument: land grant and other . . . .

Fence or field line . . . .

Although this number of symbols appears to cover a fairly good spectrum of subject matter, there are, in fact, several times this many. Other sources of such information are available through the Department of Commerce, U. S. Coast and Geodetic Survey, U. S. Army, U. S. Geological Survey, National Ocean and Atmospheric Administration, U.S. Department of Interior etc.

Answers

Exercise 2-1.

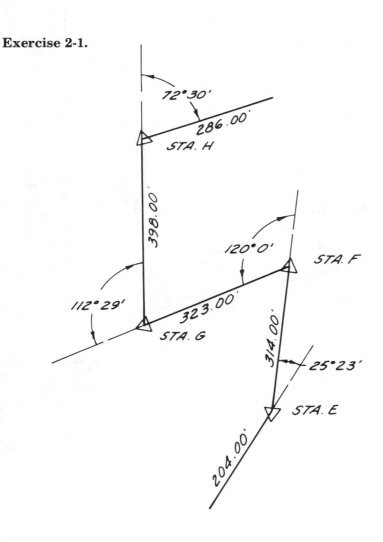

Exercise 2-3.

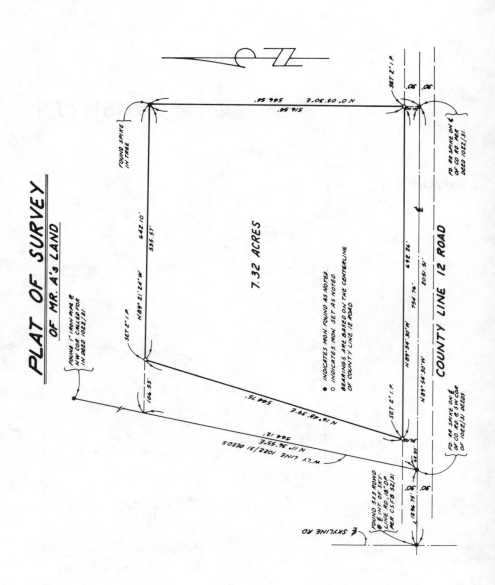

Exercise 3-1.

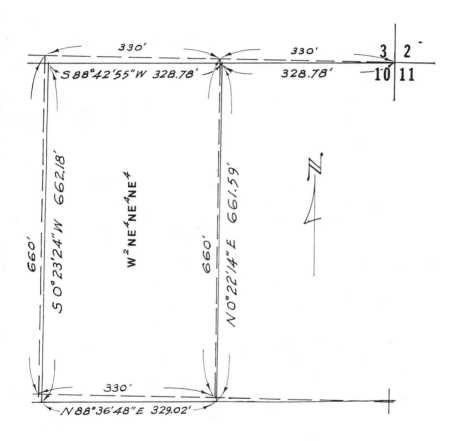

Exercise 3-2.

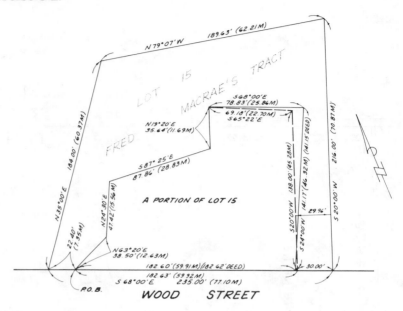

Exercise 6-1.

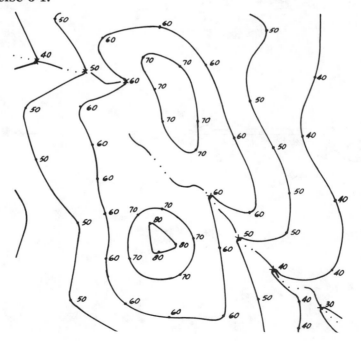

Exercise 6-2.

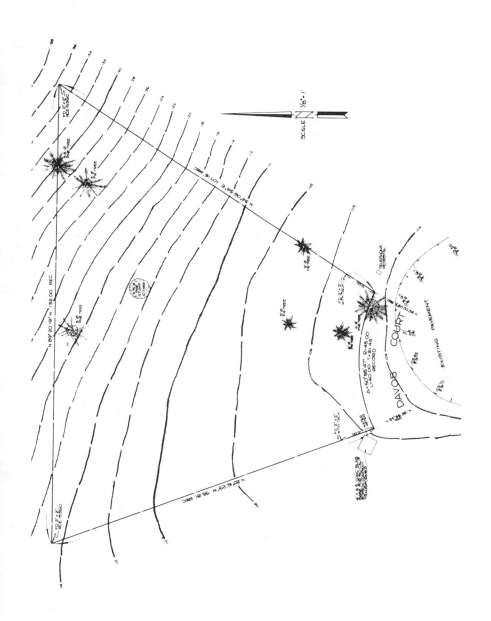

Exercise 6-3.

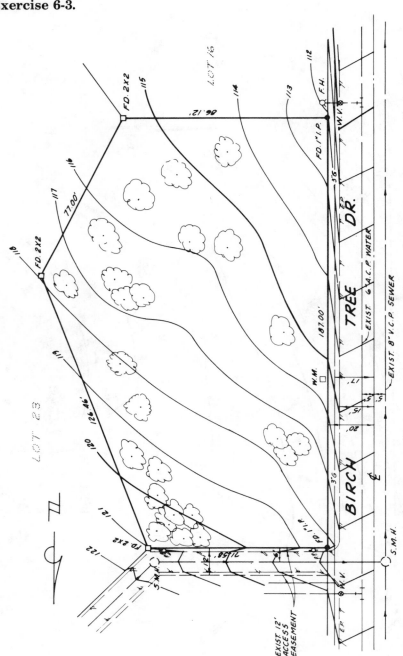

Exercise 7-1.

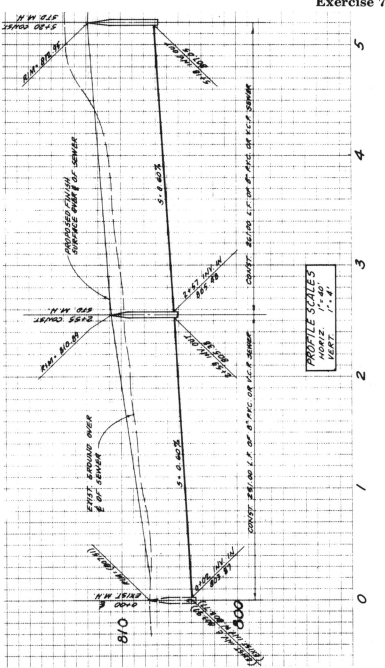

Exercise 8-1.

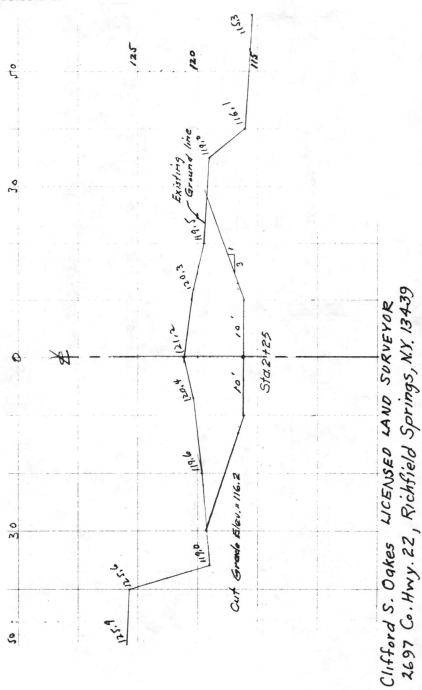

Exercise 8-2.

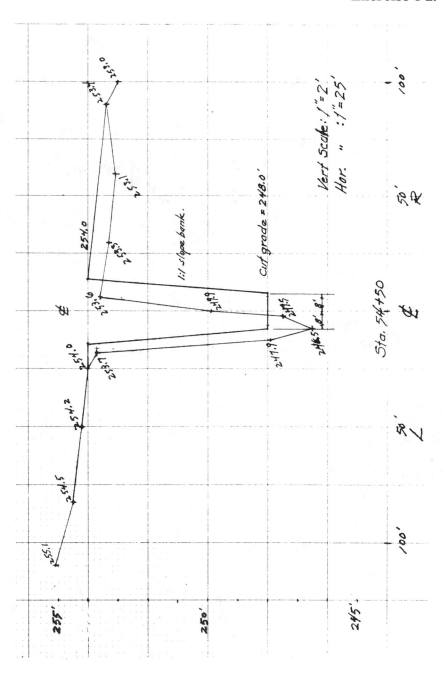

Exercise 9-1.

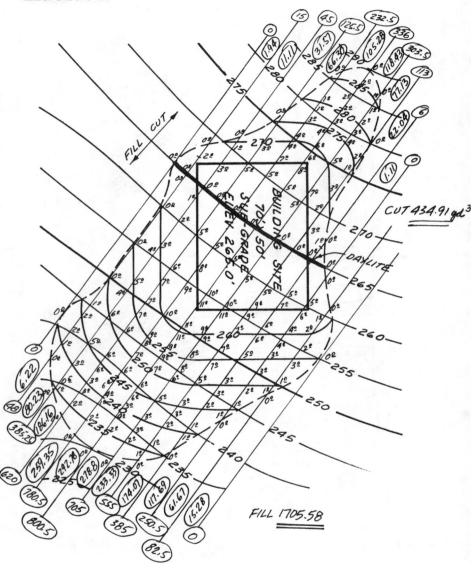

It is a recognized fact that volumes of dirt calculated by cross-sections and by planimeter will not result in exactly the same figures. For example, if the cross-sections are taken too far apart, they will not include enough of the undulations between them to give the true picture.

The work sheet used for cross-sectioning is shown above.

In this problem based on Figure 9-18, we have:

	Total Fill	Total Cut
By planimeter:	1700.80 yds³	425.70 yds³
By cross-section:	1705.58 yds³	434.91 yds³

This is reasonably close and makes a good check.
The work sheet used for planimetering is shown below.

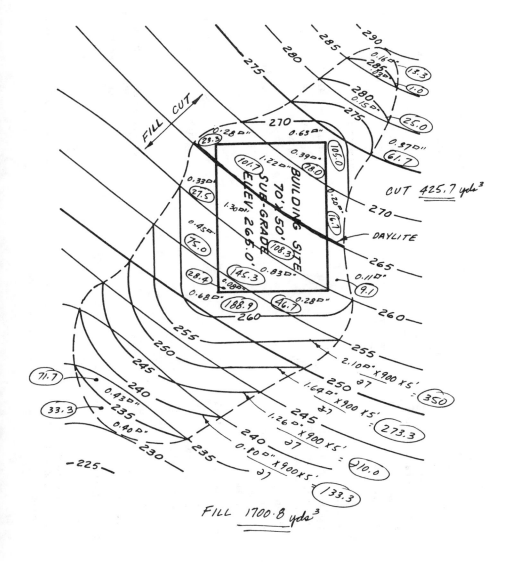

Exercise 9-2.

Cube:
Surface — $S = 6l^2$
Volume — $V = l^3$

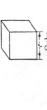

Cube

Rectangular Parallelopiped:
Surface — $S = 2(lb + bh = hl)$
Volume — $V = bh$

Parallelopiped

Prism or Cylinder:
Surface — $S = $ perimeter of a right section times length of the lateral edge.
Volume — $V = $ area of the base times altitude at right angles to base.

Prism or Cylinder

Pyramid or Cone (symmetrical):
Surface — $S = $ ½ perimeter of base times slant height.

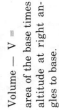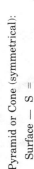

Pyramid or Cone

Volume — $V = $ ⅓ area of base times altitude.

Frustrum of Pyramid or Cone (symmetrical):
Surface — $S = $ ½ sum of perimeters of bases times slant height.

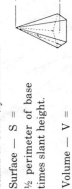

Frustrum of Pyramid or Cone

Frustrum of Pyramid or Cone (symmetrical):

Volume — $V = $
$⅓(A_1 + A_2 + \sqrt{A_1 \times A_2})h$
where A_1 and A_2 are the two bases of the frustrum.

Frustrum of Pyramid or Cone

Prismatoid:

Volume — $V = $
$\frac{1}{6}(A_1 + A_2 + 4A_m)h$
where A_1 and A_2 are the two bases and A_m is the area of the mid cross-section parallel with the bases

Prismatoid

Spheres:

Total area — $S = 4\pi^2$ or πD^2
Total Volume — $V = \frac{4}{3}\pi R^3$ or $\frac{1}{6}\pi D^3$

Area & Vol. of Sphere

Volume of a Segment with one base —
$PV = \frac{1}{3}\pi h^2(3R - h)$
or $\frac{1}{6}\pi Rh(3R^2 + h^2)$

Seg. Vol. of Sphere

Zone area — $ZA = 2\pi Rh$ or πDh

Zone Area

Exercise 10-1.

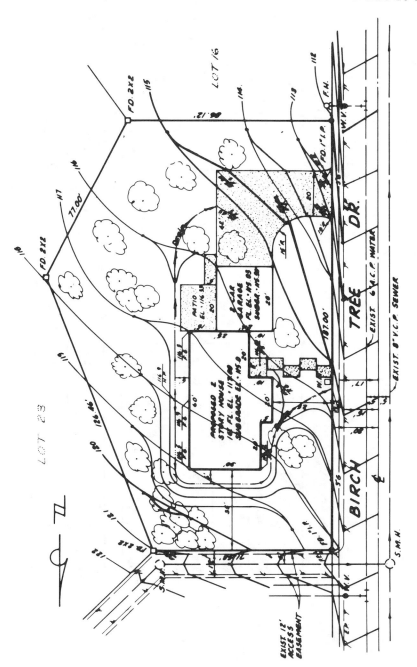

Exercise 11-1.

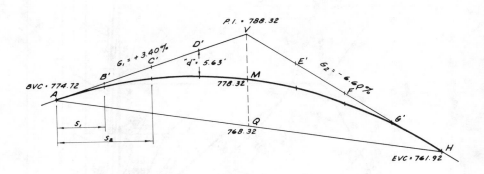

$$Q = \frac{A+H}{2} \qquad m = \frac{V+Q}{2}$$

PROFILE STATION	PROB. STA.	ELEV. ON TANGENT	S	S²	VERT OFFSET	ELEV. ON CURVE	1ST DIFF.	2ND DIFF
6+00	A	774.72	0	0	0	774.72		
							−2.77	
7+00	B'	778.12	1/4	1/16	−0.63	777.49		1.24
							−1.53	
8+00	C'	781.52	2/4	1/4	−2.50	779.02		1.26
							−0.27	
9+00	D'	784.92	3/4	9/16	−5.63	779.29		1.24
							+0.97	
10+00	V	788.32	1	1	−10.00	778.32		1.26
							+2.23	
11+00	E'	781.72	3/4	9/16	−5.63	776.09		1.24
							+3.47	
12+00	F'	775.12	2/4	1/4	−2.50	772.62		1.26
							+4.73	
13+00	G'	768.52	1/4	1/16	−0.63	767.89		1.24
							+5.97	
14+00	H	761.92	0	0	0	761.92		

Exercise 12-1.

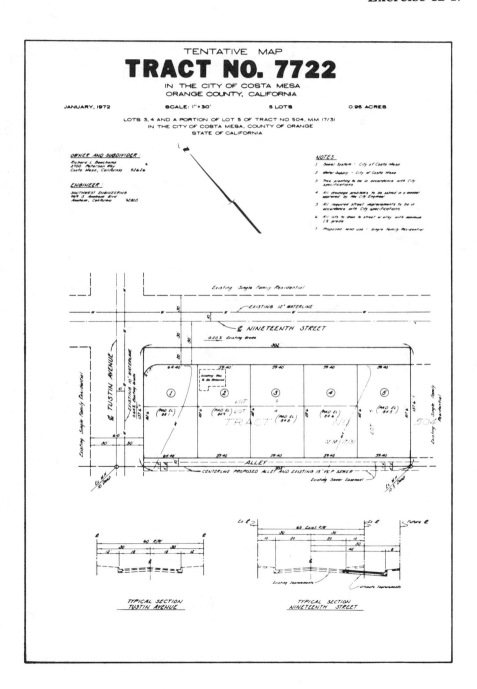

Exercise 12-1.

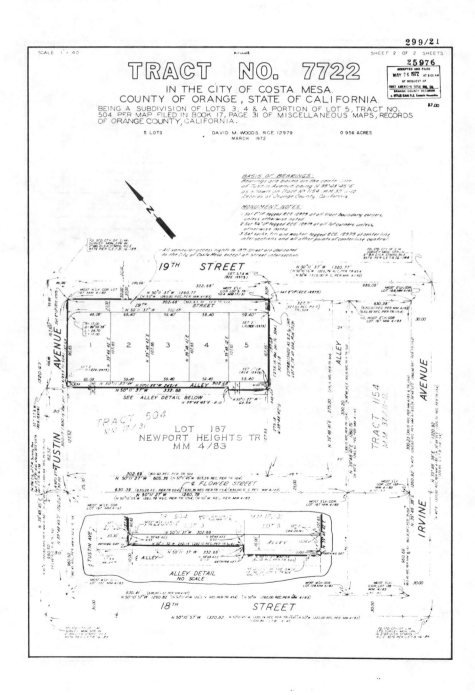

Exercise 12-2.

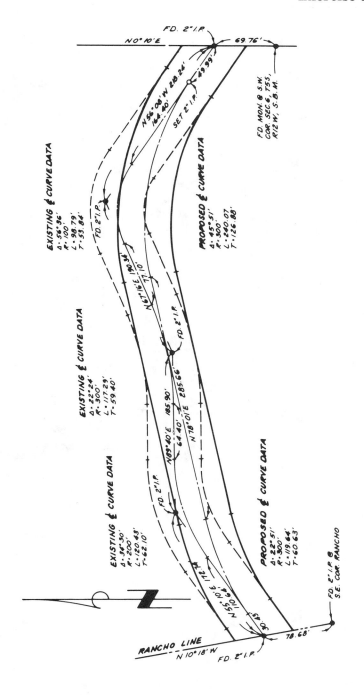

Exercise 12-3.

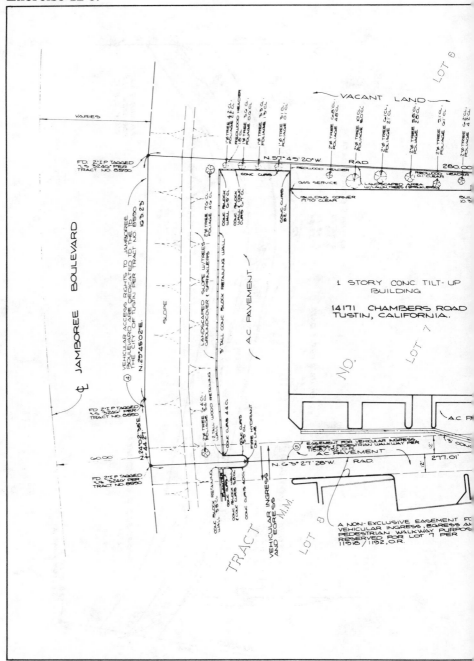

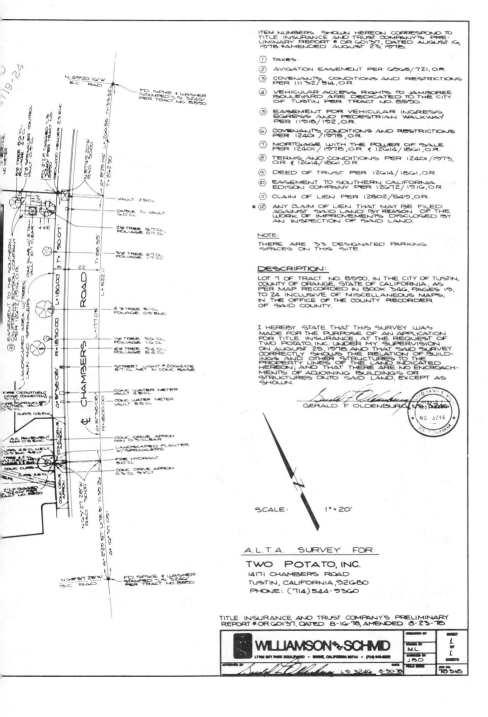

ITEM NUMBERS SHOWN HEREON CORRESPOND TO TITLE INSURANCE AND TRUST COMPANY'S PRELIMINARY REPORT # OR 60137, DATED AUGUST 16, 1978. AMENDED AUGUST 23, 1978.

1. TAXES.
2. AVIGATION EASEMENT PER 6965/721, O.R.
3. COVENANTS, CONDITIONS AND RESTRICTIONS PER 11152/514, O.R.
4. VEHICULAR ACCESS RIGHTS TO JAMBOREE BOULEVARD ARE DEDICATED TO THE CITY OF TUSTIN PER TRACT NO. 8590.
5. EASEMENT FOR VEHICULAR INGRESS, EGRESS AND PEDESTRIAN WALKWAY PER 11918/192, O.R.
6. COVENANTS, CONDITIONS AND RESTRICTIONS PER 12401/1975, O.R.
7. MORTGAGE WITH THE POWER OF SALE PER 12401/1978, O.R. & 12614/1861, O.R.
8. TERMS AND CONDITIONS PER 12401/1973, O.R. & 12614/1861, O.R.
9. DEED OF TRUST PER 12614/1861, O.R.
10. EASEMENT TO SOUTHERN CALIFORNIA EDISON COMPANY PER 12672/1916, O.R.
11. CLAIM OF LIEN PER 12802/549, O.R.
* 12. ANY CLAIM OF LIEN THAT MAY BE FILED AGAINST SAID LAND BY REASON OF THE WORK OF IMPROVEMENTS DISCLOSED BY AN INSPECTION OF SAID LAND.

NOTE:
THERE ARE 33 DESIGNATED PARKING SPACES ON THIS SITE.

DESCRIPTION:
LOT 7 OF TRACT NO. 8590, IN THE CITY OF TUSTIN, COUNTY OF ORANGE, STATE OF CALIFORNIA, AS PER MAP RECORDED IN BOOK 346, PAGES 19 TO 24 INCLUSIVE OF MISCELLANEOUS MAPS, IN THE OFFICE OF THE COUNTY RECORDER OF SAID COUNTY.

I HEREBY STATE THAT THIS SURVEY WAS MADE FOR THE PURPOSE OF AN APPLICATION FOR TITLE INSURANCE, AT THE REQUEST OF TWO POTATO, INC. UNDER MY SUPERVISION ON AUGUST 25, 1978 AND THAT SAID SURVEY CORRECTLY SHOWS THE RELATION OF BUILDINGS AND OTHER STRUCTURES TO THE PROPERTY LINES OF THE LAND INDICATED HEREON; AND THAT THERE ARE NO ENCROACHMENTS OF ADJOINING BUILDINGS OR STRUCTURES ONTO SAID LAND, EXCEPT AS SHOWN.

GERALD F. OLDENBURG, LS 3246

SCALE: 1" = 20'

A.L.T.A. SURVEY FOR

TWO POTATO, INC.
14171 CHAMBERS ROAD
TUSTIN, CALIFORNIA 92680
PHONE: (714) 544-9360

TITLE INSURANCE AND TRUST COMPANY'S PRELIMINARY REPORT # OR 60137, DATED 8-16-78, AMENDED 8-23-78

WILLIAMSON & SCHMID

LS 3246 8-30-78

Exercise 12-4.

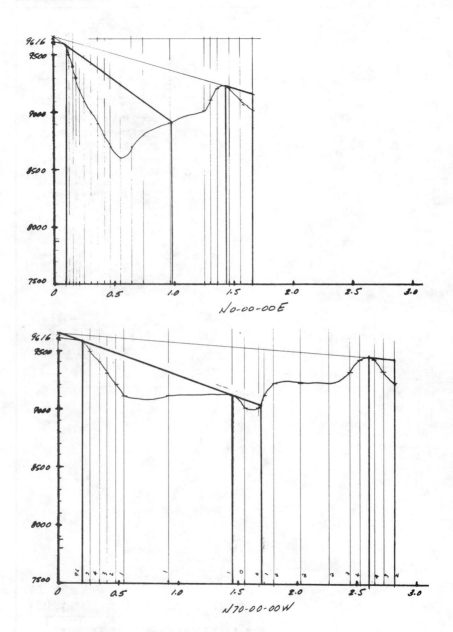

N 0-00-00 E

N 70-00-00 W

2 of the profiles used for final answer

Exercise 12-4.

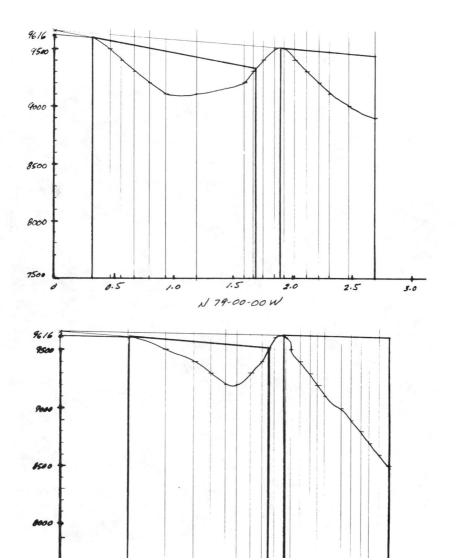

2 of the profiles used for final answer

Exercise 12-4.

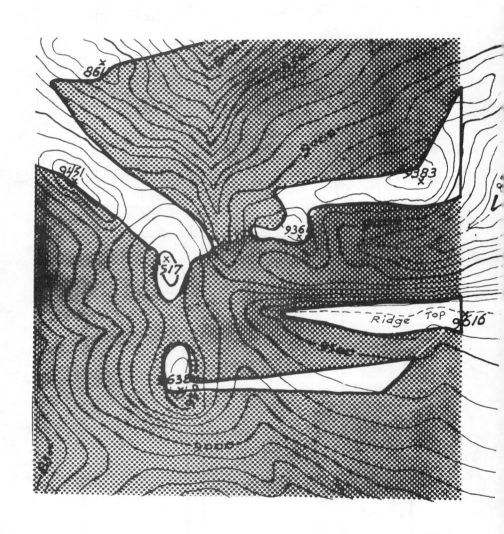

Index